THE NEW PLAGUES:
Pandemics and Poverty in a Globalized World

Since time immemorial, infectious diseases and epidemics have decimated entire populations, triggered mass migrations, and decided the outcome of wars. The threat of infectious diseases still hangs over us today – now more than ever due to the quickening pace of globalization. Infectious diseases affect every facet of our lives. Besides being the focus of research and medicine, they shape society and culture and have a significant economic and political impact. This book describes the threat of infectious diseases in our globalized world and discusses ways and means of dealing with it.

Stefan H. E. Kaufmann is a Professor of Microbiology and Immunology, and Founding Director of the Max Planck Institute for Infection Biology in Berlin.

Our addresses on the Internet:
www.the-sustainability-project.com
www.forum-fuer-verantwortung.de
[English version available]

THE NEW PLAGUES:
Pandemics and Poverty in a Globalized World

STEFAN H. E. KAUFMANN
With the assistance of Susan Schädlich

Translated by Michael Capone

Klaus Wiegandt, General Editor

HAUS PUBLISHING

First published in Great Britain in 2009 by
Haus Publishing Ltd
70 Cadogan Place
London SW1X 9AH
www.hauspublishing.com

Originally published as: *Wächst die Seuchengefahr: Globale Epidemien und Armut: Strategien zur Seucheneindämmung in einer vernetzten Welt* by Stefan H E Kaufmann

A CIP catalogue record for this book
is available from the British Library

ISBN 978-1-906598-13-6

Typeset in Sabon by MacGuru Ltd
Printed in Dubai by Oriental Press

Contents

Editor's Foreword

Sustainability Project

Sales of the German-language edition of this series have exceeded all expectations. The positive media response has been encouraging, too. Both of these positive responses demonstrate that the series addresses the right topics in a language that is easily understood by the general reader. The combination of thematic breadth and scientifically astute, yet generally accessible writing, is particularly important as I believe it to be a vital prerequisite for smoothing the way to a sustainable society by turning knowledge into action. After all, I am not a scientist myself; my background is in business.

A few months ago, shortly after the first volumes had been published, we received suggestions from neighboring countries in Europe recommending that an English-language edition would reach a far larger readership. Books dealing with global challenges, they said, require global action brought about by informed debate amongst as large an audience as possible. When delegates from India, China, and Pakistan voiced similar concerns at an international conference my mind was made up. Dedicated individuals such as Lester R. Brown and Jonathan Porritt deserve credit for bringing the concept of sustainability to the attention of the general public, I am convinced that this series can give the discourse about sustainability something new.

Two years have passed since I wrote the foreword to the initial German edition. During this time, unsustainable developments on our planet have come to our attention in ever more dramatic ways. The price of oil has nearly tripled; the value of industrial metals has risen exponentially and, quite unexpectedly, the costs of staple foods such as corn, rice, and wheat have reached all-time highs. Around the globe, people are increasingly concerned that the pressure caused by these drastic price increases will lead to serious destabilization in China, India, Indonesia, Vietnam, and Malaysia, the world's key developing regions.

The frequency and intensity of natural disasters brought on by global warming has continued to increase. Many regions of our Earth are experiencing prolonged droughts, with subsequent shortages of drinking water and the destruction of entire harvests. In other parts of the world, typhoons and hurricanes are causing massive flooding and inflicting immeasurable suffering.

The turbulence in the world's financial markets, triggered by the US sub-prime mortgage crisis, has only added to these woes. It has affected every country and made clear just how unscrupulous and sometimes irresponsible speculation has become in today's financial world. The expectation of exorbitant short-term rates of return on capital investments led to complex and obscure financial engineering. Coupled with a reckless willingness to take risks everyone involved seemingly lost track of the situation. How else can blue chip companies incur multi-billion dollar losses? If central banks had not come to the rescue with dramatic steps to back up their currencies, the world's economy would have collapsed. It was only in these circumstances that the use of public monies could be justified. It is therefore imperative to prevent a repeat of speculation with short-term capital on such a gigantic scale.

Taken together, these developments have at least significantly improved the readiness for a debate on sustainability. Many more are now aware that our wasteful use of natural resources and energy have serious consequences, and not only for future generations.

Two years ago, who would have dared to hope that WalMart, the world's largest retailer, would initiate a dialog about sustainability with its customers and promise to put the results into practice? Who would have considered it possible that CNN would start a series "Going Green?" Every day, more and more businesses worldwide announce that they are putting the topic of sustainability at the core of their strategic considerations. Let us use this momentum to try and make sure that these positive developments are not a flash in the pan, but a solid part of our necessary discourse within civic society.

However, we cannot achieve sustainable development through a multitude of individual adjustments. We are facing the challenge of critical fundamental questioning of our lifestyle and consumption and patterns of production. We must grapple with the complexity of the entire earth system in a forward-looking and precautionary manner, and not focus solely on topics such as energy and climate change.

The authors of these twelve books examine the consequences of our destructive interference in the Earth ecosystem from different perspectives. They point out that we still have plenty of opportunities to shape a sustainable future. If we want to achieve this, however, it is imperative that we use the information we have as a basis for systematic action, guided by the principles of sustainable development. If the step from knowledge to action is not only to be taken, but also to succeed, we need to offer comprehensive education to all, with the foundation in early childhood. The central issues of the future must be anchored firmly in school curricula, and no university student should be permitted to graduate without having completed a general course on sustainable development. Everyday opportunities for action must be made clear to us all – young and old. Only then can we begin to think critically about our lifestyles and make positive changes in the direction of sustainability. We need to show the business community the way to sustainable development via a responsible attitude to consumption, and become active within our sphere of influence as opinion leaders.

For this reason, my foundation *Forum für Verantwortung*, the ASKO EUROPA-FOUNDATION, and the European Academy Otzenhausen have joined forces to produce educational materials on the future of the Earth to accompany the twelve books developed at the renowned Wuppertal Institute for Climate, Environment and Energy. We are setting up an extensive program of seminars, and the initial results are very promising. The success of our initiative "Encouraging Sustainability," which has now been awarded the status of an official project of the UN Decade "Education for Sustainable Development," confirms the public's great interest in, and demand for, well-founded information.

I would like to thank the authors for their additional effort to update all their information and put the contents of their original volumes in a more global context. My special thanks goes to the translators, who submitted themselves to a strict timetable, and to Annette Maas for coordinating the Sustainability Project. I am grateful for the expert editorial advice of Amy Irvine and the Haus Publishing editorial team for not losing track of the "3600-page-work."

Taking Action – Out of Insight and Responsibility

"We were on our way to becoming gods, supreme beings who could create a second world, using the natural world only as building blocks for our new creation."

This warning by the psychoanalyst and social philosopher Erich Fromm is to be found in *To Have or to Be?* (1976). It aptly expresses the dilemma in which we find ourselves as a result of our scientific-technical orientation.

The original intention of submitting to nature in order to make use of it ("knowledge is power") evolved into subjugating nature in order to exploit it. We have left the earlier successful path with its many advances and are now on the wrong track, a path of danger with incalculable risks. The greatest danger stems from the unshakable faith of the overwhelming majority of politicians and business leaders in unlimited economic growth which, together with limitless technological innovation, is supposed to provide solutions to all the challenges of the present and the future.

For decades now, scientists have been warning of this collision course with nature. As early as 1983, the United Nations founded the World Commission on Environment and Development which published the Brundtland Report in 1987. Under the title *Our Common Future*, it presented a concept that could save mankind from catastrophe and help to find the way back to a responsible way of life, the concept of long-term environmentally sustainable use of resources. "Sustainability," as used in the Brundtland Report, means "development that meets the needs of the present without compromising the ability of future generations to meet their own needs."

Despite many efforts, this guiding principle for ecologically, economically, and socially sustainable action has unfortunately not yet become the reality it can, indeed must, become. I believe the reason for this is that civil societies have not yet been sufficiently informed and mobilized.

Forum für Verantwortung

Against this background, and in the light of ever more warnings and scientific results, I decided to take on a societal responsibility with my foundation. I would like to contribute to the expansion of public discourse about sustainable development which is absolutely essential. It is my desire to provide a large number of people with facts and contextual knowledge on the subject of sustainability, and to show alternative options for future action.

After all, the principle of "sustainable development" alone is insufficient to change current patterns of living and economic practices. It does provide some orientation, but it has to be negotiated in concrete terms within society and then implemented in patterns of behavior. A democratic society seriously seeking to reorient itself towards future viability must rely on critical, creative individuals capable of both discussion and action. For this reason, life-long learning, from childhood to old age, is a necessary precondition for realizing sustainable development. The practical implementation of the ecological, economic, and social goals of a sustainability strategy in economic policy requires people able to reflect, innovate and recognize potentials for structural change and learn to use them in the best interests of society.

It is not enough for individuals to be merely "concerned." On the contrary, it is necessary to understand the scientific background and interconnections in order to have access to them and be able to develop them in discussions that lead in the right direction. Only in this way can the ability to make appropriate judgments emerge, and this is a prerequisite for responsible action.

The essential condition for this is presentation of both the facts and the theories within whose framework possible courses of action are visible in a manner that is both appropriate to the subject matter and comprehensible. Then, people will be able to use them to guide their personal behavior.

In order to move towards this goal, I asked renowned scientists to present in a generally understandable way the state of research and the possible options on twelve important topics in the area of sustainable development in the series "*Forum für Verantwortung*." All those involved in this project are in agreement that there is no alternative to a united path of all societies towards sustainability:

- *Our Planet: How Much More Can Earth Take?* (Jill Jäger)
- *Energy: The World's Race for Resources in the 21st Century* (Hermann-Joseph Wagner)
- *Our Threatened Oceans* (Stefan Rahmstorf and Katherine Richardson)
- *Water Resources: Efficient, Sustainable and Equitable Use* (Wolfram Mauser)
- *The Earth: Natural Resources and Human Intervention* (Friedrich Schmidt-Bleek)
- *Overcrowded World? Global Population and International Migration* (Rainer Münz and Albert F. Reiterer)
- *Feeding the Planet: Environmental Protection through Sustainable Agriculture* (Klaus Hahlbrock)
- *Costing the Earth? Perspectives on Sustainable Development* (Bernd Meyer)

The public debate

What gives me the courage to carry out this project and the optimism that I will reach civil societies in this way, and possibly provide an impetus for change?

For one thing, I have observed that, because of the number and severity of natural disasters in recent years, people have become more sensitive concerning questions of how we treat the Earth. For another, there are scarcely any books on the market that cover in language comprehensible to civil society the broad spectrum of comprehensive sustainable development in an integrated manner.

When I began to structure my ideas and the prerequisites for a public discourse on sustainability in 2004, I could not foresee that by the time the first books of the series were published, the general public would have come to perceive at least climate change and energy as topics of great concern. I believe this occurred especially as a result of the following events:

First, the United States witnessed the devastation of New Orleans in August 2005 by Hurricane Katrina, and the anarchy following in the wake of this disaster.

Second, in 2006, Al Gore began his information campaign on climate change and wastage of energy, culminating in his film *An*

Inconvenient Truth, which has made an impression on a wide audience of all age groups around the world.

Third, the 700-page Stern Report, commissioned by the British government, published in 2007 by the former Chief Economist of the World Bank Nicholas Stern in collaboration with other economists, was a wake-up call for politicians and business leaders alike. This report makes clear how extensive the damage to the global economy will be if we continue with "business as usual" and do not take vigorous steps to halt climate change. At the same time, the report demonstrates that we could finance countermeasures for just one-tenth of the cost of the probable damage, and could limit average global warming to 2° C – if we only took action.

Fourth, the most recent IPCC report, published in early 2007, was met by especially intense media interest, and therefore also received considerable public attention. It laid bare as never before how serious the situation is, and called for drastic action against climate change.

Last, but not least, the exceptional commitment of a number of billionaires such as Bill Gates, Warren Buffett, George Soros, and Richard Branson as well as Bill Clinton's work to "save the world" is impressing people around the globe and deserves mention here.

An important task for the authors of our twelve-volume series was to provide appropriate steps towards sustainable development in their particular subject area. In this context, we must always be aware that successful transition to this type of economic, ecological, and social development on our planet cannot succeed immediately, but will require many decades. Today, there are still no sure formulae for the most successful long-term path. A large number of scientists and even more innovative entrepreneurs and managers will have to use their creativity and dynamism to solve the great challenges. Nonetheless, even today, we can discern the first clear goals we must reach in order to avert a looming catastrophe. And billions of consumers around the world can use their daily purchasing decisions to help both ease and significantly accelerate the economy's transition to sustainable development – provided the political framework is there. In addition, from a global perspective, billions of citizens have the opportunity to mark out the political "guide rails" in a democratic way via their parliaments.

The most important insight currently shared by the scientific, political, and economic communities is that our resource-intensive Western model of prosperity (enjoyed today by one billion people) cannot be extended to another five billion or, by 2050, at least eight billion people. That would go far beyond the biophysical capacity of the planet. This realization is not in dispute. At issue, however, are the consequences we need to draw from it.

If we want to avoid serious conflicts between nations, the industrialized countries must reduce their consumption of resources by more than the developing and threshold countries increase theirs. In the future, all countries must achieve the same level of consumption. Only then will we be able to create the necessary ecological room for maneuver in order to ensure an appropriate level of prosperity for developing and threshold countries.

To avoid a dramatic loss of prosperity in the West during this long-term process of adaptation, the transition from high to low resource use, that is, to an ecological market economy, must be set in motion quickly.

On the other hand, the threshold and developing countries must commit themselves to getting their population growth under control within the foreseeable future. The twenty-year Programme of Action adopted by the United Nations International Conference on Population and Development in Cairo in 1994 must be implemented with stronger support from the industrialized nations.

If humankind does not succeed in drastically improving resource and energy efficiency and reducing population growth in a sustainable manner – we should remind ourselves of the United Nations forecast that population growth will come to a halt only at the end of this century, with a world population of eleven to twelve billion – then we run the real risk of developing eco-dictatorships. In the words of Ernst Ulrich von Weizsäcker: "States will be sorely tempted to ration limited resources, to micromanage economic activity, and in the interest of the environment to specify from above what citizens may or may not do. 'Quality-of-life' experts might define in an authoritarian way what kind of needs people are permitted to satisfy." (*Earth Politics*, 1989, in English translation: 1994).

It is time

It is time for us to take stock in a fundamental and critical way. We, the public, must decide what kind of future we want. Progress and quality of life is not dependent on year-by-year growth in per capita income alone, nor do we need inexorably growing amounts of goods to satisfy our needs. The short-term goals of our economy, such as maximizing profits and accumulating capital, are major obstacles to sustainable development. We should go back to a more decentralized economy and reduce world trade and the waste of energy associated with it in a targeted fashion. If resources and energy were to cost their "true" prices, the global process of rationalization and labor displacement will be reversed, because cost pressure will be shifted to the areas of materials and energy.

The path to sustainability requires enormous technological innovations. But not everything that is technologically possible has to be put into practice. We should not strive to place all areas of our lives under the dictates of the economic system. Making justice and fairness a reality for everyone is not only a moral and ethical imperative, but is also the most important means of securing world peace in the long term. For this reason, it is essential to place the political relationship between states and peoples on a new basis, a basis with which everyone can identify, not only the most powerful. Without common principles of global governance, sustainability cannot become a reality in any of the fields discussed in this series.

And finally, we must ask whether we humans have the right to reproduce to such an extent that we may reach a population of eleven to twelve billion by the end of this century, laying claim to every square centimeter of our Earth and restricting and destroying the habitats and way of life of all other species to an ever greater degree.

Our future is not predetermined. We ourselves shape it by our actions. We can continue as before, but if we do so, we will put ourselves in the biophysical straitjacket of nature, with possibly disastrous political implications, by the middle of this century. But we also have the opportunity to create a fairer and more viable future for ourselves and for future generations. This requires the commitment of everyone on our planet.

Klaus Wiegandt

Summer 2008

Preface

Hardly a year goes by without some infectious disease dominating the headlines. This is usually followed by hysterical activities that gradually peter out. We find the unknown and unexpected especially threatening. Reading names such as BSE, SARS, and H5N1 in newspapers, we might be forgiven for thinking that infectious diseases claim just a few hundred lives. But these are just the tip of the iceberg. What the press fails to mention is that around 50,000 people die from infectious diseases every day.

When I was asked if I would write a book about the threat of infectious disease for the Forum for Responsibility series, I did not immediately jump at the chance. The task seemed far too daunting on top of all my other commitments. In the end I agreed, because I believe that something urgently needs to be done to contain infectious diseases. In retrospect the project proved enjoyable, and I learned a great deal more than I thought I would. As I delved into the subject, one thing became abundantly clear: something needs to be done – and soon!

Infectious diseases affect every facet of our lives. They are the focus of research and medicine; they shape our society and culture; and they have a significant economic and political impact. Within this network, they are both cause and consequence. What has been lacking so far is an attempt to view the various aspects from different angles and to unravel the complex web of interdependencies. The purpose of this book is to do just that – not in the technical language of the scientist but in a way that will enable as many people as possible to form their own opinions about the globalization of infectious diseases in the modern, networked world.

I was fortunate to have had the unstinting support of a reliable and helpful group of people. I wish to thank Dr. Mary Louise Grossman for her research for this book, Diane Schad for preparing the instructive illustrations, Souraya Sibaei for her unflagging help writing the manuscript and her sound research, and Susan Schädlich for her competent and always stimulating assistance. Eva Köster at Fischer Verlag and Anette Maas at the Forum for Responsibility lent their support to the project with great commitment. I wish to thank my colleague Professor Klaus Hahlbrock for his patience in persuading me to write this book. I would also like to extend my special thanks to the Forum for Responsibility, particularly to Klaus Wiegandt, who kindled my fascination for the project and generously supported me. Many colleagues read parts of the manuscript. In particular, I wish to thank Professor Martin Grobusch, Professor Frank Kirchhoff, Professor Peter Kremsner, Professor Klaus Magdorf, Professor Kai Matuschewski, and Professor Richard Lucius. When I accepted the book project I knew that I would be writing it mainly in my free time. I thank Elke, my wife and my sons Moritz and Felix for their incredible patience and forbearance while I devoted so much time to this book instead of to them. It wasn't the first time.

1 Introduction

Knowledge is not enough; it must also be applied.
Desire is not enough; it must be acted upon.

Johann Wolfgang von Goethe

Microorganisms colonized the Earth three billion years ago. Today the planet teems with 500,000 to one million different species of bacteria and around 5000 species of viruses, most of which have yet to be characterized. The majority are harmless and do not bother us. We know of around 1500 microorganisms that cause infectious diseases in humans. Most of them have thankfully remained rarities. Nevertheless, transmissible diseases are responsible for one-third to one-quarter of all deaths today. Again and again they have decimated entire populations, triggered mass migrations, and decided the outcome of wars.

As recently as World War II, infectious diseases were able to spread almost unopposed. Although vaccines had been developed to prevent smallpox, diphtheria, and tetanus, people had to rely for the most part on hygiene, disinfection, and sterilization measures. The 1950s saw the advent of anti-infectives, drugs for preventing and treating infections. In addition, effective vaccines were developed against several major viral diseases: measles, mumps, rubella (German measles), and polio. The second breakthrough came with the discovery of antibiotics, which specifically attack bacteria. Between the 1950s and the 1970s the treatment of bacterial infectious diseases became part of routine medical practice.

Today we are witnessing a resurgence of transmissible diseases – and we only have ourselves to blame. We have shown ourselves incapable of supplying poor countries with vaccines and antibiotics in sufficient quantities. Moreover, pathogens, i.e. disease-causing microorganisms, are constantly developing new forms of resistance to our antibiotics and are consequently becoming difficult or even impossible to fight. Finally, our lifestyle is helping to create highly favorable conditions for new pathogens to develop and flourish.

In the following I invite you to learn more about the importance of transmissible diseases for our future. This book deals with scientific, medical, economic, social, and political aspects, and presents evidence that infectious diseases have not become any less of a threat to humanity. On the contrary, pathogens are proving to be the major winners in the globalization of our world. Microorganisms are able to respond to changes and adapt to new situations faster than any other form of life. Globalization, a growing chasm between developing and industrialized nations, numerous catastrophes and crises, and not least the industrialization of our food sources, and our conquest of the last remnants of unspoiled nature are opening up unprecedented opportunities for pathogens. More than ever before people are in constant interaction with each other. A pathogen that affects one person can spread to the entire world within one to two days. As a result, a new infectious disease, even if it originates in a remote corner of the world, can quickly spread to threaten the entire globe.

On the other hand, humans have more means to defend themselves against pathogens than ever before. We understand precisely how pathogens emerge and cause diseases; we are able to detect outbreaks promptly; and we have at our disposal an arsenal of diagnostic agents, vaccines, and drugs. What we lack, however, is a determined will to act, to apply the available resources, and to develop new strategies. During the five minutes or so it has taken you to read this introduction, fifty people have become infected with the human immunodeficiency virus (HIV). In addition, tuberculosis has claimed seventeen lives, and malaria has ended the lives of ten children. Had we exploited our resources effectively, many of those people would still be alive.

I will discuss the threat of infectious diseases in our globalized world and strategies for controlling them. I hope that at the end of this book you will agree with me that infectious diseases – more than ever before – have a significant impact on society, economics, and politics and are not just the consequence but also the cause of many of the world's problems.

This book will first look at the strategies used by microorganisms and the types of defenses the body mounts to fight them. I will then discuss those transmissible diseases that are of greatest concern at the moment and the means we have available for preventing and treating them. I will also address the question as to why today's world is a rich breeding ground for old and new pathogens. Finally, I will propose solutions for preventing new outbreaks of infectious diseases and for combating existing ones.

I would like to mention one thing from the outset: reports of successes have unfortunately become the exception in recent decades. Gradually we are running out of options. Drugs and vaccines generate too little profit for the market to provide sufficient incentives for their development and production. New, innovative approaches are called for. We have the knowledge and the financial means to effect a sea change, but we need to act and we need to act now!

2 The Invaders

On any possible, reasonable or fair criterion, bacteria are – and always have been – the dominant forms of life on Earth.

Stephen Jay Gould

2.1 Introduction

The dread we have of infectious diseases owes much to the fact that they are transmissible. Our fears are probably also fed by the fact that we cannot prevent them from spreading. Pathogens have many ways of jumping from one person to another: the virus of the immunodeficiency disease AIDS is transmitted mainly through sexual intercourse, and it is for this reason that condoms are able to provide protection. Certain mosquitoes transmit malaria, and impregnated bed nets are still the most effective means of preventing the disease. The causative agents of bacillary dysentery, cholera, and other diarrheal diseases are transmitted from person to person via contaminated water or food; a great deal can be achieved simply by boiling drinking water and peeling and washing fruits and vegetables, even if this is sometimes inconvenient when traveling. Microbes, such as those that cause tuberculosis, are swirled through the air in tiny droplets – a mode of transmission against which there is little protection. Such aerosols, as they are known, are produced not only by affected individuals coughing and sneezing; air-conditioning systems can also spread microbes that dwell in cooling-water reservoirs. That was the cause of the first outbreak of Legionnaires' disease in 1976. The disease is caused by *Legionella* bacteria and manifests itself as severe pneumonia accompanied by fever. In our latitudes Legionnaires' disease captures the public's attention whenever the bacteria are found in public swimming pools, which often then have to be closed for weeks while they are disinfected.

In other cases infections are transmitted via direct skin contact (contagious diseases), via contact with infectious materials such as blood (blood transfusions or sharing of needles by drug addicts), or via bodily fluids during sexual intercourse. Other possible routes of infection are via sewage, foods, or infected pets and wild animals. Animals may carry pathogens even if they themselves appear healthy. Healthy carriers among the human population pose a similarly deceptive threat. These individuals carry a pathogen and spread it without manifesting the disease themselves. Finally, insects such as mosquitoes, fleas, and ticks figure prominently as carriers of many infectious diseases and in this role are known as vectors.

Blocking the transmission path of pathogens is often the best way to prevent or contain infectious diseases.

Infectious diseases are caused by microscopic organisms, i.e. bacteria, viruses, fungi, and protozoa, or by larger parasites. We have recently discovered that even simple proteins known as prions can transmit diseases. These pathogens, discovered just a few years ago, cause spongy dissolution of the brain, e.g. in mad cow disease in cattle (also known as BSE for bovine spongiform encephalopathy), scrapie in sheep, and Creutzfeld-Jakob disease in humans.

I will describe the most important pathogens and the diseases they cause later. In the following I will describe the smallest organisms and their fascinating survival strategies, which are important for understanding the nature of infectious diseases.

2.2 Bacteria

Bacteria are often only a fraction of the diameter of a human hair in size. For example, *Escherichia coli*, a typical intestinal microbe, measures just a few micrometers across – about one hundredth the thickness of a human hair. Bacteria occur in many shapes and most retain their typical form for life (see Fig. 1). Thus the bacteria that cause anthrax and tuberculosis are rods; the typical pus-forming bacteria, staphylococci and streptococci, are spheres; and the bacterium that causes cholera is comma-shaped. The causative agent of tick-borne Lyme disease (borreliosis) and the syphilis bacterium are corkscrew-shaped.

Many bacteria require oxygen to live. They are therefore referred to as aerobic microbes. Other bacteria are able to thrive without oxygen (facultative anaerobes), while for others oxygen acts as a deadly poison (obligatory anaerobes). The latter include most intestinal bacteria and those that multiply in deep wounds, such as the bacteria that cause tetanus.

Bacteria are independent organisms. Although they lack a true cellular nucleus, bacteria, like higher life forms, carry genetic information about their survival strategies encoded in DNA. Bacteria multiply asexually by undergoing division. No genetic material from paternal and maternal lines is combined or swapped. Instead, bacteria replicate their DNA before dividing and then allocate one copy to each of the two resulting daughter cells. The progeny are therefore genetically identical to the mother (or father) cell. Researchers therefore refer to a group of such cells as a bacterial clone.

A BACTERIUM

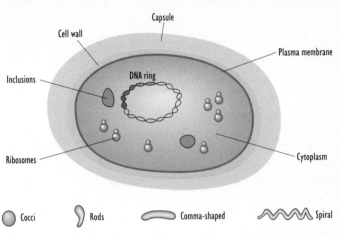

Figure I   Structure of a bacterium

Bacteria have a rigid cell wall surrounding the plasma membrane and are
sometimes enveloped by a capsule. The cell wall and capsule protect the
bacterium from environmental influences and the body's defense mechanisms.
The genetic material is contained in a DNA ring, not in a set of chromosomes.
The cytoplasm contains, among other things, inclusions and ribosomes, in
which the RNA is translated into protein. Bacteria occur in various shapes,
which are stable for a given microbe. The most important forms are spherical
cocci, oblong rods, comma-shaped, and spiral-shaped forms.

Many bacteria multiply very rapidly, i.e. once every 30 minutes
to one hour. A few, such as the notoriously slow-breeding bacillus
that causes tuberculosis, take much longer, i.e. 12 to 24 hours. The
bacterium that causes leprosy is even slower, taking two weeks to
divide. The ability to divide rapidly has an important advantage:
working flat out, the DNA copier keeps making small errors.
As a result, not every bacterial great-great-great grandchild is
perfectly identical to its ancestor down to the last DNA building
block. Such random mutations may have no effect whatsoever, or

they may manifest themselves as incremental changes that confer a survival advantage or disadvantage on the bacterium. Those variants that are better adapted to the environment quickly become dominant. Even if a dramatic DNA mutation were to cause 99.999% of the bacteria to perish, the surviving microbes would be able to repopulate the niche very rapidly. Mutation and selection are the mainsprings of evolution. Over the course of millions of years they have enabled bacteria to colonize almost every conceivable habitat on Earth. The same mechanisms help microbes develop counterstrategies, for example to antibiotics, and to evolve drug-resistant strains.

Despite their asexual mode of propagation, bacteria nevertheless swap genetic material with each other. There are two principal ways in which this occurs. The first route is via viruses that infect bacteria. The group of viruses known as bacteriophages consists – like all viruses – chiefly of genetic material. When they move from one bacterium to another they often transport DNA segments, such as genes for antibiotic resistance. Secondly, besides their actual genetic material, some bacteria also contain separate rings of DNA floating inside them. These structures, known as plasmids, can also be swapped between bacteria. Because this gene exchange takes place between two coexisting microbes within one bacterial generation, scientists refer to the process as horizontal gene exchange. This mechanism is particularly important in the case of infectious diseases, as it allows microbes to transfer entire pathogenic traits in one fell swoop. Thus, harmless intestinal bacteria such as *Escherichia coli* may receive information from diarrhea-causing microbes such as salmonellae or shigellae with the result that they too are then able to cause diarrhea. Conversely, a harmless *Escherichia coli* bacterium that long ago developed resistance to antibiotics may transfer that ability to a diarrhea-causing microbe that had previously been susceptible to antibiotics.

Because bacteria possess an independent metabolism that differs markedly from our own, antibiotics are able to kill them selectively without disrupting our own metabolic processes. These valuable drugs only became available in the 1950s and since then have saved millions of lives. However, the weapon is slowly becoming blunt precisely because bacteria are able to adapt so rapidly and develop evasive strategies that render them insensitive to antibiotic treatment. Antibiotics and antibiotic resistance are dealt with in detail in a separate chapter.

2.3 Viruses

"A virus is a piece of bad news wrapped in protein." With this sentence the British Nobel laureate in medicine Peter Medawar described the nature of viruses in a nutshell.

Viruses are about ten times smaller than bacteria (i.e. 0.1 micrometers or 1000 times thinner than a human hair). They are not independent organisms but rather typical parasites. Essentially, they consist of a shell enclosing the information required for their production in a fragment of genetic material (DNA or RNA). Viruses contain either DNA or RNA but never both at the same time. In higher organisms, by contrast, genetic information is encoded in DNA. RNA serves as a messenger that reads the blueprint and translates it into protein according to a code. Viruses lack at least part of this synthesis apparatus. Consequently, they must rely upon more complex cells to replicate. Having invaded a cell, they commandeer its metabolic machinery and use it for their own devices (see Fig. 2). The infected cell then replicates the virus and usually bursts in the end, releasing hoards of viral offspring.

In viruses that contain RNA instead of DNA there is an

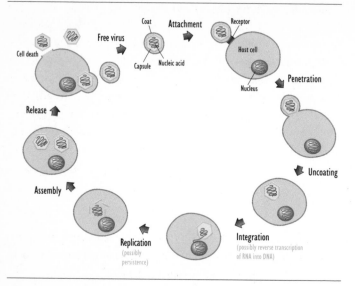

Figure 2   Virus replication

Viruses dock to the host cell via specific receptors, which also determine the
range of host and organ specificity of the virus. Once the virus has penetrated
into the cell, its capsule opens and the viral DNA is integrated into the host
genome. In the case of RNA viruses the information carrier must first be
reverse-transcribed into DNA. The virus can then replicate immediately or
persist for a long time in a sort of latency stage and become active later.
Copying of the viral RNA or DNA is followed by assembly of the capsule and
possibly the external coat. The free virus is then ready to infect other host cells.

intermediate step, because the cell must reverse-transcribe the
RNA into DNA before it can replicate. This process often results
in reading errors. As in the case of bacteria, the errors may be
advantageous in some cases. For example, such reading errors
allow HIV to adopt new forms continuously and evade the
body's immune response.

Because viruses are much smaller than bacteria, it only
became possible to visualize them and study them in detail with

the invention of the electron microscope, and effective drugs against viruses were not developed until the 1990s. Antibiotics and other conventional antibacterial drugs are ineffective against viruses. Thus, the widespread practice of treating non-bacterial tonsillitis with antibiotics is futile, since it is likely to be caused by viruses. Antibiotics are a sensible option only in bacterial infections. In streptococcal angina, for example, they can be life-saving.

2.4 Protozoa

Protozoans are single-celled organisms that bear many similarities to higher organisms, including humans. Most protozoans that can potentially cause disease are found in tropical and sub-tropical areas of Africa, Asia, and South America. They include the parasite *Plasmodium*, which causes malaria; trypanosomes, which cause Chagas disease and sleeping sickness; and *Leishmania*, the agent of leishmaniasis (kala-azar). Other protozoans, such as *Toxoplasma*, which is transmitted mainly by cats, and *Giardia*, which causes diarrhea, also occur in our climes. Many protozoans are transmitted by insects. Because the metabolism of protozoans is quite similar to our own, drugs that affect them also harm the body's cells. Consequently, our arsenal of antiprotozoan drugs is limited.

2.5 Fungi and worms

Some fungi and worms also cause disease. Many fungi are opportunists. Though abundant, they cause little harm to healthy individuals. However, in the presence of a compromised immune system fungi can cause serious diseases that are often extremely difficult to treat. This is a huge problem especially for AIDS patients.

Worm infestations have been largely eradicated in our latitudes, but one in three people in the world still plays host to at least one worm species. Those most frequently affected are schoolchildren in the tropics. The importance of worm infestations is reflected in the fact that the body's immune system has evolved specific anti-worm mechanisms which differ markedly from those directed against bacteria and viruses.

2.6 Prions

Prions are the smallest infectious agents known. In recent years they have completely upset our understanding of disease. Prions, which are simply incorrectly folded proteins, are certainly not organisms. For many years they have been known to be the causative agents of Creutzfeldt-Jakob disease. However, with the outbreak of bovine spongiform encephalitis (BSE), or mad cow disease, as it is popularly known, they suddenly captured the headlines in the early 1990s. The outbreak, which ultimately claimed the lives of hundreds of thousands of cattle, was caused by human activities: we forced these herbivorous animals into cannibalism by feeding them ground bones and processed fats from other animals. These cost-saving measures used in industrial animal breeding culminated in an economic catastrophe and sowed the seeds of persistent doubt and uncertainty in the minds of consumers. The BSE crisis cost an estimated 3.5 billion euros. However, despite initial fears, a catastrophic epidemic in humans did not occur. Up to June 2008 only 208 (including 167 in the UK, 23 in France, 3 in the USA and Spain) cases of new-variant Creutzfeldt-Jakob disease, thought to be the human form of BSE, had been reported. Prions will therefore not feature prominently in this book.

3 The Defenders

> Blood is a very special juice.
>
> Johann Wolfgang von Goethe

3.1 Introduction

The immune response is one of the body's most underrated functions: it has no problem distinguishing between billions of different structures. Metaphorically speaking, it would be perfectly capable of precisely distinguishing each and every one of the 6.5 billion human beings who inhabit the Earth. And by using the lock-and-key principle it is capable of producing just as many structures of its own to capture each and every one of these countless extraneous structures. The immune system performs this task so effectively that we are generally completely unaware of how the body combats thousands of potential pathogens every day. Our immune system is such an extraordinarily powerful weapon that it is difficult to avoid using military analogies when describing it. I have always tried to do so, but have yet to find a more apt alternative. At the same time, the weapon that is our immune system is subject to rigid control in order to prevent excessive or misdirected reactions and resulting damage to the body itself such as occurs, for example, in allergies and in diseases in which the immune system attacks the tissues of its own body.

The ancient Romans used the Latin term *immunis* to refer to freedom from the obligation to pay tax. Today we use the word "immunity" to refer to freedom from infectious diseases. The outstanding characteristics of immunity are its specificity and its phenomenally precise memory. Without really understanding why, most people know that a child who has had mumps will not catch it again, but may well catch measles or rubella (German measles). Even in Ancient Greece the historian Thucydides observed that people who had suffered an infectious disease did not contract the same disease again. And King Dionysius reported that he regularly took minute amounts of poison at breakfast time in order to protect himself against attempts to poison him. It is doubtful whether this would have induced an immune response, though it may have "trained" his liver enzymes to break poisons down more rapidly – just as a person who enjoys a daily glass of wine increases his tolerance to alcohol.

3.2 Making it difficult to penetrate the lines

Let's look at how the human body prevents pathogens from gaining entry. The first barrier is formed by a number of natural resistance mechanisms. The skin forms the first obstacle to intruders (Fig. 3). Most would-be intruders can enter the body only via open wounds or else through bodily orifices that are lined by mucous membrane. However, when passing through mucous membranes they are confronted by other defense mechanisms. The airways, for example, are lined by a cell layer that bears a coating of tiny hair-like structures known as cilia that transport intruding dust particles and microorganisms back up towards their point of entry into the body. In the intestine, waves of muscular contraction known as peristalsis drive the intestinal

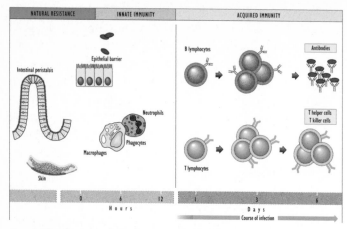

Figure 3   **The first steps in resistance to infection**

In order to gain entry to a host, a harmful microorganism must first overcome the
barriers of natural resistance. The skin is very well protected against would-be
intruders, which are unable to penetrate the skin unless it is damaged. The
epithelial cells that line the airways are coated with tiny hair-like structures known
as cilia that can transport microorganisms back out of the body. In the intestine,
waves of muscular contraction known as peristalsis ensure that the intestinal
contents including microorganisms are eliminated from the body in the stool.
Any microorganisms that manage to overcome the barriers of natural resistance
are then attacked by various cells and factors of innate immunity. An important
role in this respect is played by macrophages and neutrophilic granulocytes, which
engulf intruding microorganisms. Only after a few days does acquired immunity
begin its targeted attack on the intruders. It does this firstly by means of specific
antibodies, which are produced by B lymphocytes, and secondly by means of T
helper cells and T killer cells, which develop from precursor lymphocytes

contents including microorganisms out of the body. In this way
humans eliminate millions of bacteria with every bowel movement. In a similar way, the flow of urine continuously flushes
microorganisms out of the body. In addition, certain beneficial
bacteria establish themselves naturally on many body surfaces.
Thus, the bacterial flora that is normally present in the intestine
or in the mouth and pharynx prevents colonization by harmful

microorganisms. To use an analogy, any intruder that arrives in these areas is entering an overcrowded train in which all the seats and most of the standing room are already occupied. It will have great difficulty finding any reasonably comfortable position, i.e. anywhere to get an infection going.

As important as all these mechanisms are, they have little to do with immunity in the narrow sense. Immunity as such comes into play only after a pathogen has overcome all these external barriers.

3.3 Behind the barriers

The immune system stands on two pillars: one pillar is innate immunity, the other is acquired immunity (Fig. 3). The innate part recognizes pathogens near their point of entry into the body and mobilizes the initial defense forces. To continue with the analogy of a military engagement, these initial defense forces can be seen as foot soldiers who initiate the defense without any sophisticated military equipment. The macrophage and granu-locyte units eliminate many pathogens simply by engulfing (ingesting) them. Immune cells reinforce each other by secret-ing various soluble substances: for example, "complement" dis-solves bacteria, while interferons block viral reproduction. Using these strategies, the innate immune system acts effectively, albeit imprecisely. Sooner or later a wily pathogen will overrun these foot soldiers.

However, the innate immune system keeps in close contact with the next line of defense, namely the special forces advanc-ing from the rear that constitute the acquired immune system. The foot soldiers take advantage of the first counter-attack to spy on the enemy: via a number of relatively nonspecific, but extraordinarily effective, mechanisms they find out what type of enemy they are facing. Are the attackers bacteria that could cause an acute infection? Are they viruses that could initiate a chronic infection? Or perhaps fungi, protozoa, or worms? The foot soldiers pass this information on to the acquired immune system. At a sort of command level within the latter a decision is made as to which specific defense battalions need to be mobilized. To follow the analogy, is it the army, the navy, or the air force that is required?

The acquired immune system undertakes the tasks that we normally associate with the term "immunity". It prepares a targeted counter-attack known as the specific immune response and possesses a memory that is located in cells and factors present in the blood, the lymph, and inflamed tissue. White blood cells known as lymphocytes recognize all structures that are foreign to the body. These are known as antigens. In order to repulse these, the special unit adopts a dual approach: firstly, it designs an appropriate defensive protein, or specific antibody, for each such invader; and secondly, it dispatches specialized immune cells to meet the invaders.

Antibodies are Y-shaped proteins that fit like a key in a lock. They attach themselves to the invaders and in this way neutralize toxins or initiate resistance to bacteria or viruses. They are produced by B lymphocytes, also known simply as B cells.

Various classes of antibody, or immunoglobulin, are distinguished and for the sake of simplicity referred to by means of letters: IgA, IgD, IgG, IgM, and IgE. Natural antibodies of the first line of resistance belong to the IgM class. IgD antibodies sit on the surface of B cells and in a sense "frisk" the intruders. Antibodies of the IgG class form a specific counter-attacking unit and are the most important immune substances present in the blood. IgE antibodies are deployed especially in worm infestations and mediate symptoms in allergic reactions. Finally, the mucous membranes of the intestine, mouth, throat, lungs, urinary bladder, and vagina – i.e. the principal entry portals for microorganisms – have an antibody class of their own, namely IgA.

Against pathogens that park themselves in host cells, however, antibodies are powerless. They fail to recognize intruders that are hiding in body cells. Infected cells nevertheless inform the immune system of the presence within themselves of invaders by fitting their surface out with new antigens as a sort of distress signal. By means of this ploy they attract another type of immune cell, namely T lymphocytes, or T cells. Among other functions, these cells are able to destroy infected cells and thereby stop production of the virus. They can also activate macrophages so that these destroy pathogens hidden within them. Activation of macrophages is carried out by T helper cells, while another group of T helper cells controls antibody production. Destruction of infected cells is accomplished by T killer cells.

Like the foot soldiers, our nonspecific defenders, T helper cells also make use of a number of messenger substances. These are known as cytokines or interleukins. By means of these hormone-like proteins the T helper cells exchange information with their target cells over shorter or longer distances. The cytokine network is complex, and often a number of these messenger substances have to act together to mobilize a particular function.

T helper cells of type 1 (Th1 cells) activate macrophages which thereupon kill bacteria and protozoa. T helper cells of type 2 (Th2 cells) stimulate antibody production and combat worm infestations. Th1 and Th2 cells activate various functions by producing messenger substances. The most important messenger substances produced by Th1 cells are interferon gamma, the most potent activator of macrophages, and interleukin 2, which activates T killer cells. Interleukin 4 and interleukin 5, which are produced by Th2 cells, coordinate antibody production and resistance to worm infestations (Fig. 4).

In many diseases the special unit itself is attacked. AIDS is a well-known example of this. The lethal immunodeficiency virus uses recognition patterns on the surface of T cells as a docking site. It binds specifically to CD4 molecules on the surface of T helper cells and via these gains entry to the immune cells. Other cells bear related receptors: T killer cells, for example, have CD8 molecules in their cell membrane. In the laboratory medical scientists also make use of these molecules for diagnostic purposes. For example, the ratio of CD4 to CD8 molecules on T cells can be used to predict the course of AIDS.

In many cases the acquired immune system is too specialized to be able to combat pathogens by itself. Instead, it often devises the strategy and issues specific directives but then once again calls upon the innate immune system – in our analogy, the foot soldiers – to attack. At the same time the acquired immune system develops an immunologic memory. The command level records the important details of all attackers and in the event of a second invasion is able to intervene more promptly and with greater precision (Fig. 5).

In general the immune system takes about 5–15 days after first contact with an intruder to get fully going. For that reason many acute infectious diseases show an improvement after one to two weeks. Similarly, vaccines take some weeks to provide reliable protection. On second contact the immune response gets into full swing earlier and often becomes effective within a few days. This is why we contract childhood diseases only once.

One last important task remains after the immune system has overcome the enemy: the defense forces that have been deployed must be withdrawn before their weapons are directed against

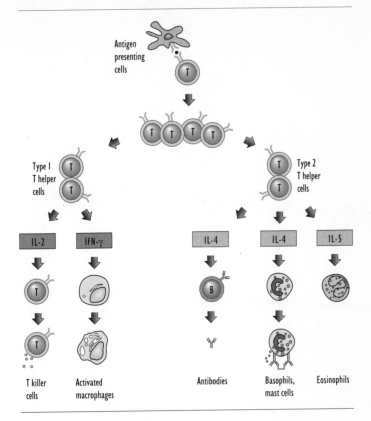

Figure 4 The system of T helper cells

When T lymphocytes recognize their specific antigen they reproduce and develop a number of different functions. T helper cells stimulate other cells by means of soluble mediators. Two main groups of T helper cells are distinguished: T helper cells of type 1 produce, among other things, the soluble mediators interleukin 2 (IL-2) and interferon gamma (IL-γ) and in this way activate T killer cells and macrophages; T helper cells of type 2 produce, among other things, interleukin 4 (IL-4) and interleukin 5 (IL-5), which stimulate B lymphocytes to produce antibodies and activate basophils, mast cells, and eosinophils for protection against worms. However, basophils, mast cells, and eosinophils also cause allergic reactions.

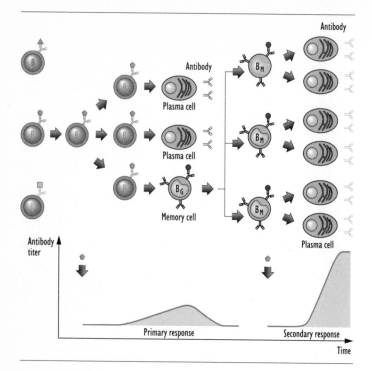

Figure 5   Development of immunologic memory

When B lymphocytes react to their specific antigen by means of their receptors –
i.e. antibodies expressed on their surface – they replicate and develop into two
different types of cells: plasma cells, which produce antibodies that attack the
pathogens directly; and memory cells, which go into action only if the same
pathogen attacks a second time. If this occurs, the memory B (B_M) cells are
immediately ready to produce specific antibodies. The secondary response is
therefore considerably more effective than the primary response. A very similar
process occurs with T lymphocytes. The formation of memory cells is an
important precondition for vaccine-induced immunity.

their own body. Suppressor mechanisms are responsible for this. If these fail to function correctly, an autoimmune disease can result.

3.4 Help for the defense: vaccines

In the summer of 1885 Louis Pasteur (1822–1895) inoculated Joseph Meister, a nine-year-old boy from Alsace who had become infected with rabies, with an extract of the spinal cord of rabid rabbits. Though Pasteur had very little understanding of immunology at that time, his attempt to induce immunity saved the boy's life. Without knowing exactly how, Pasteur had effectively stimulated the boy's immune system. The mechanism by which this occurred is the same with all active vaccines. As much as present-day vaccines may differ from each other, they all have one thing in common: they dangle a supposed attacker in front of the immune system and thereby provoke it into action. They are all more or less direct copies of a pathogenic organism or of disease-inducing factors present in a pathogen, the only difference being that they are essentially harmless. The immune system is unable to distinguish between the copy and the original, i.e. between the vaccine and the pathogenic organism. Ideally, therefore, immunization induces an effective reaction against the pathogen. The information required for this reaction is retained in the body – and in the event of a real attack by the organism concerned can be accessed much more rapidly than previously.

Traditional vaccines can be divided into the following groups:

Just as Pasteur's vaccine for Joseph Meister contained real rabies viruses that the scientist had weakened, *live vaccines*, i.e. vaccines consisting of pathogenic organisms that have been attenuated (rendered harmless), are still used today. Live vaccines are used, for example, to immunize infants against measles, mumps, and rubella. The smallpox vaccine is also based on the same principle, while the tuberculosis vaccine BCG (for *bacille Calmette-Guérin*, after the two French scientists who developed it) was produced by attenuating the organism responsible for bovine tuberculosis. Attenuated vaccines tend to be very effective, but the possibility cannot be completely excluded that the living microorganisms may, by a process of back-mutation, reacquire pathogenic properties, though in practice strict controls make this extremely unlikely. In addition, live vaccines contain many components that can cause side effects.

The second group is that of *inactivated vaccines*. Here too whole pathogens are used, but only after being killed. Influenza vaccines are inactivated vaccines. Since the inoculated viruses are dead, they cannot cause any infectious disease. Influenza vaccines consist of a cocktail of influenza viruses that are considered likely to cause the next seasonal influenza epidemic.

Inactivated vaccines contain many components that by themselves do not contribute to immunity and may be harmful. More and more work is therefore being conducted on *subunit vaccines*, which ideally contain only the antigens that induce an immune response. To this end the organisms are split and subunits isolated. Unfortunately, however, purified antigens do not generally induce a strong immune response by themselves. Subunit vaccines therefore require the addition of substances – known as adjuvants – that "stoke" the immune reaction. The hepatitis B vaccine is an example of an adjuvant-containing subunit vaccine. It consists of an artificially (e.g. in yeasts) produced surface protein of the hepatitis virus plus an adjuvant, in most cases aluminum hydroxide.

A subgroup of subunit vaccines is formed by *toxoid vaccines* directed against microbial toxins. In this case inactivated toxins (poisonous substances) that have lost their harmful properties are inoculated. The immune system remembers these. The best-known examples are the diphtheria and tetanus vaccines. Another subgroup of subunit vaccines is that of *conjugate vaccines*. These are used against bacterial pathogens that the immune system recognizes on the basis of polysaccharide (sugar) components present on their surface. This is the case, for example, with *Haemophilus influenzae* type b, a common cause of meningitis (inflammation of the membranes that surround the brain) in young children. Our immune system produces antibodies against these polysaccharides, but keeps no record of the carbohydrates concerned, i.e. no immunologic memory results from contact with them. However, protein fragments that have been artificially attached – conjugated – to these polysaccharide molecules can act as an aid to memory. By conjugating such components to tetanus or diphtheria toxoid it is possible to produce vaccines that provide protection not only against tetanus and diphtheria, but also against *H. influenzae*. The vaccines that are used to protect against pneumococci and meningococci, typical pathogens of the ear, nose, and throat regions, are also conjugate vaccines.

Different pathogens require different types of vaccine that differ in terms of their purity and effectiveness. Whereas live vaccines and inactivated vaccines contain many components that can potentially cause harmful effects, subunit vaccines, and to an even greater extent vaccines that consist of a single antigen, are far more pure and are essentially free of harmful components. On the other hand, increasing purity is often accompanied by diminishing effectiveness. By itself, a vaccine made of purified components induces less immunity than does a live vaccine that can survive in the body for some time. As a remedy to this problem, pure antigens are combined with highly active adjuvants whenever possible. At present, however, few such substances are available. The development of adjuvants will occupy an important place in the future of vaccine research.

No effective vaccine prevents infection from occurring at the outset; instead, vaccines prevent infections from causing illness. They do this mostly by activating B lymphocytes, which in response to administration of a vaccine start to produce antibodies. In other words, all vaccines that have proved successful to date owe their effectiveness principally to humoral immunity. This indicates the primary objective of new vaccine development, namely to develop vaccines for those diseases in which T lymphocytes act as bearers of immunity. This is not to say that such vaccines should stimulate only T lymphocytes; rather, they should activate both arms of the immune system, i.e. antibody production and T cells. Unfortunately, we do not yet know how best to stimulate the T cells that are needed for immunity.

A second problem confronting the development of new vaccines is the breadth of the range of pathogens to be targeted. On the one hand, clinicians and researchers would naturally wish the immunity that is induced by vaccines to be optimally effective against and specific to a particular pathogen. On the other hand, it is clear that this is not always the best approach. For example, it would be preferable to possess a single vaccine for all pneumococci that was directed against an antigen shared by all strains of the organism. The same is true of influenza: at present we rely on specific vaccines that have to be composed anew each year. How much better it would be if we could develop a single vaccine directed against a conserved antigen.

This is a perfectly feasible objective, and here again T lymphocytes would be the center of attention, since we know that it is they that recognize infected cells on the basis of conserved structures shared by all influenza viruses. Pathogens such as HIV that undergo constant change in the host would likewise be far better combated by a vaccine directed against structures that do not change.

At least equally desirable would be vaccines which, rather than preventing the development of illness, prevented infection. In the case of acute infectious diseases with a short interval between infection and (prevention of) onset of illness, this might not be so important, however in chronic infections the length of the incubation period can be crucially important: in HIV, for example, the immune system eventually breaks down altogether. In such diseases it is difficult to see how protection induced by a vaccine could control the pathogen in the long term. Better would be a vaccine that blocked or completely eliminated the pathogen immediately after it gained entry to and established itself in the body. An immunity that merely checked the proliferation of HIV would become ineffective as AIDS developed. A vaccine that induced this kind of immunity could only delay, not prevent, the onset of disease.

It is to be feared that this kind of situation now exists with tuberculosis. Though in some cases an infected individual can keep the bacteria under control for a long time and thereby delay the onset of illness, a vaccine that prevented illness from occurring would be a welcome advance. What would happen, however, if the immune system were to collapse some time later, for instance as a result of simultaneous HIV infection? In that event even the best vaccine would be of no avail. Here again, a vaccine that prevented infection from occurring in the first place would be far more useful.

AIDS and tuberculosis are two global infectious diseases that pose great challenges for vaccine research. Malaria is another.

And a vaccine that could remove our fear of an H5N1 pandemic in humans would be extremely helpful.

Passive immunization, which was introduced by Emil Behring (1854–1917), is even making something of a comeback at present. In view of the worrying increase in multiresistant nosocomial pathogens, a number of biotechnology companies have decided to develop antibodies as therapeutic agents. Except that in place of antisera obtained from experimentally infected animals such as were used in Behring's time to treat diphtheria and tetanus, scientists are now developing humanized monoclonal antibodies that our immune system does not regard as foreign and that attack certain pathogens in a highly specific fashion.

3.5 Misdirected immunity: allergy and autoimmune diseases

Notwithstanding the excellent work that it does, our immune system can also get things wrong. Every immune response also has the potential to cause harm. In autoimmune diseases and allergies the body's protective mechanisms themselves cause illness. Statistics show that such misprogramming of immunity is extremely common. Thus, 20% of the US population and 15% of the UK population suffer from allergies, while in Australia an astounding 45% of the population suffer from allergies and in Germany up to 20% of adults and over 10% of children are troubled by allergies.

In the USA rheumatoid arthritis afflicts more than two million people, psoriasis up to 7.5 million, and multiple sclerosis 400,000, giving a total of 10 million people with these three autoimmune diseases. In the UK more than 500,000 people suffer from rheumatoid arthritis, 1.2 million from psoriasis, and almost 90,000 from multiple sclerosis. In Australia more than 600,000 people are troubled by these same three autoimmune disorders (183,000 rheumatoid arthritis, 400,000 psoriasis, and 28,000 multiple sclerosis). According to the German Society for Autoimmune Diseases, autoimmune diseases such as inflammatory joint diseases, multiple sclerosis, and psoriasis affect more than five million people in Germany, Switzerland, and Austria.

Allergies and asthma are the result of a misdirected immune response to parasites. Autoimmune diseases are often brought about by a misdirected reaction to bacterial or viral pathogens and are therefore triggered by Th1 cells. In this process each arm of the T helper cell response acts as a control on, and therefore influences, the other. Th1 cells of type 1 directed against bacteria and viruses suppress the immune response to worms, while the immune response to worms suppresses the response to bacteria and viruses. It is believed that as a result, an underdeveloped Th1 antibacterial response favors the development of Th2-mediated allergy (see Fig. 3). Proponents of the "hygiene hypothesis" point to such an association. According to this hypothesis, the reason for the increase in the number of people with allergies is that in the antibacterial environment of modern Western civilization the immune system only rarely comes into contact with real threats. Less genuine (Th1-mediated) immune response to bacteria, so it is concluded, results in more Th2-mediated allergy.

Allergy: taking a sledgehammer to crack a nut

In allergies the immune system appears to go crazy. It mounts an elaborate attack on foreign substances that are actually completely harmless – and in so doing damages the body. In principle, any substance can trigger an allergy – from grass pollen and latex molecules to metals and animal products. Depending on the type of allergy concerned, the body produces antibodies on first contact with the substance. On second contact these antibodies recognize the allergen and fight it. Depending on the type of antibody involved, reactions can occur either immediately or within a few hours. They range from a rash and itch to dangerous swelling of mucous membranes that can cause shortness of breath and circulatory collapse. In contact allergies, on the other hand, it is T cells that initiate the attack on the supposed intruders. For example, the immune system might try to combat a nickel-containing ring by causing eczema.

Over the past few decades epidemiologists have noticed a dramatic increase in the number of allergies in almost all industrialized countries. Older figures are difficult to obtain, however at present roughly one person in five in the UK and the USA, and one adult in five to ten in Germany, suffers from allergy. Between 13 and 24% of Germans have at one time or another been diagnosed with hay fever and about 3% have been diagnosed with asthma. In children the trend is more pronounced: every tenth child now born in Germany will develop asthma at some time in its life. The disease affects between 10 and 30% of children in most industrialized countries. In some countries up to 40% of children are troubled by allergic reactions. The three major forms of allergy are hay fever, neurodermatitis, and asthma. In the eyes of allergologists one phenomenon is particularly striking, namely the difference between the rate of allergies in the former East Germany and that in the former West Germany, and the gradual increase in the rate in the former towards that in the latter. This trend is apparent from the statistics. The only problem is that scientists are unable to explain it fully. A number of hypotheses have been put forward, mostly based on changes in living conditions, since Germans on either side of the former border scarcely differ in terms of their genetic predisposition.

Factors that are important for susceptibility to allergies probably include environmental influences during birth and in the first few years of life. If a pregnant woman comes into frequent contact with animal manure and bacterial components, the risk that her child will develop allergies is significantly less. We also know that certain microorganisms that form part of the intestinal flora reduce susceptibility to allergies. On the other hand, we have yet to identify any environmental factors that directly induce or increase susceptibility to allergies. We know that asthma attacks and other allergic reactions can be triggered or made worse by tobacco smoke, air pollution, and contact with allergens such as pollen, but these factors do not influence susceptibility as such. Allergies are thus seen to be part of a complex system that is influenced by environmental factors, infections, genetic predisposition, and the immune system.

Autoimmune diseases: immunologic hara-kiri

It is extraordinary how the immune system is able to recognize, distinguish, and react to tens of thousands of foreign substances and yet leave endogenous (i.e. the body's own) structures alone. It does this not out of ignorance, but by active tolerance. This tolerance develops during a special phase in the maturation of white blood cells. During this period lymphocytes develop tolerance for endogenous structures known as autoantigens, while any remaining potentially self-destructive cells are eliminated.

If, however, a few lymphocytes are not subjected to this control, the first step towards the development of an autoimmune disease has been taken. Nevertheless, the mere presence of such lymphocytes specialized to attack autoantigens is not necessarily significant: a few such cells are always present in the blood. And even mature lymphocytes can still be eliminated or inactivated. In other words, a sort of natural autoimmunity exists.

Nevertheless, the immune system can become dangerous. The list of autoimmune diseases is long. It ranges from organ-specific diseases such as multiple sclerosis (MS) and type 1 diabetes to nonspecific autoimmune diseases such as systemic lupus erythematosus and rheumatoid arthritis. About 5% of people in industrialized countries suffer from an autoimmune disease. MS is the most common neurologic disease of young adults in Europe and North America. In Germany alone about 100,000 people have MS. Almost 30,000 people have MS in Australia, up to 90,000 in the UK, and about 400,000 in the USA (i.e. about one person in 700). This incurable disease, in which the protective sheath that surrounds the nerves of the brain and spinal cord becomes inflamed after being attacked by T cells, progresses in bouts and is often accompanied by visual disturbances, muscular paralyses, and sensory disturbances. The first symptoms generally appear between the age of fifteen and forty years.

In type 1 diabetes the immune response is directed against autoantigens of the pancreas. The immune system destroys the insulin-producing beta cells. Most of the more than 200 million people worldwide who suffer from diabetes have type 2 diabetes, which is due above all to lack of exercise and poor diet. Over the past few years, however, the number of people with type 1 diabetes, which is an autoimmune disease, has also risen sharply, especially in adolescents. At present 600,000 of the 7.5 million diabetes sufferers in Germany have type 1 diabetes. The number of people with diabetes of either type is three million in Australia, 3.6 million in the UK, and 21 million in the USA, while the number with type 1 diabetes is almost 25,000 in Australia, more than 75,000 in the UK, and 360,000 in the USA. In systemic lupus erythematosus (SLE) the immune response is directed against, among other things, components of the cell nucleus. Antibodies bind to these universally present antigens, become deposited in vessel walls in the kidneys and skin, and thereby damage these organs. In Germany as many as 40,000 people, most of whom are young women, suffer from SLE. In the USA the figure is as high as 1.5 million while in the UK 50,000 people are afflicted by SLE. Rheumatoid arthritis, in which there is joint inflammation, tissue destruction, and breakdown of cartilage, affects about 1% of people in Germany. In the USA, between 0.1 and 1.0% of the population (2.1 million people) is affected while 0.8% suffer from rheumatoid arthritis in the UK.

Many autoimmune diseases are associated with infections. These can trigger, or at least promote the occurrence of, "immune hara-kiri" in many different ways. The simplest way is via cross-reactions. These occur when bodily structures happen to resemble certain antigens of pathogens. This causes the body to regard the structures concerned as intruders. An example of this is rheumatic fever, a disease that can occur after group A streptococcal infections such as purulent infections of the nasopharynx. Among other things, rheumatic fever can cause inflammation of the myocardium (heart muscle). The disease is due to a structural similarity between myocardial cells and streptococci. As a result of this, antibodies directed against the bacterial antigens also dock at the (confusingly similar) autoantigens. This results in inflammation. Thanks to penicillin and other potent antibiotics, the disease has become quite rare at higher latitudes. In India, however, where group A streptococcal infections often go untreated, rheumatic fever is still the most common cause of heart disease in children. More than one million people there have a history of rheumatic fever, and each year 50,000 new people contract the disease.

More commonly than via cross-reactions, however, infections cause autoimmune disease by acting as cofactors. We all have autoimmune lymphocytes in our body, however these are normally kept well under control. In infections, especially when they become chronic, the immune response is continuously stimulated. This can result in stimulation of lymphocytes that have very little or no specificity for the pathogen, but high specificity for endogenous structures. Such autoreactive lymphocytes can then reproduce in uncontrolled fashion and attack endogenous structures.

Chronic infections, in particular, pose a severe challenge to control of the immune system, since in these the processes that mobilize the immune system are operating in parallel with the processes that restrain it.

4 Coexistence of Mankind and Microbe

It is not the strongest species that survive, nor the most
intelligent, but the ones most responsive to change.

Charles Robert Darwin

4.1 Mankind

Human beings evolved several million years ago in Africa. The
last common ancestor of chimpanzees and humans lived in
Africa about seven million years ago. The genus *Homo*, to which
modern humans belong, didn't develop until much later. Experts
still argue about which archeological finds should be classified as
belonging to human beings and which not. What is not in dispute
is that the appearance of human beings was accompanied by a
quantum leap in brain development (even though the first *Homo*
to appear in Africa had the brain volume of a two-year-old
child). Brain mass appears to have been the motor that drove evolution: only thus could *Homo*, a creature whose physical attri-
butes were scarcely impressive compared to those of other great
apes or predators, have survived and spread. Anthropologists
believe that *Homo*'s dietary habits also conferred an advantage.
As well as making increasing use of stone tools, *Homo* gradually
adopted a carnivorous diet. At first, however, these humanoids
were not quite the courageous hunters that we like to imagine

them to have been. Rather, their carnivorous ways started far less heroically with the eating of carrion. On the other hand, the high protein content of meat appears to have been a precondition for the further evolution of our ancestors and in particular for the disproportionate development of their brain.

At the same time, however, this preference for a meat diet – initially, at least, in the form of carrion – was not without risk. It can be assumed that these humanoids sometimes became infected with microorganisms. In all probability they were afflicted by protozoal and worm diseases and by viral and bacterial infections.

Major epidemics, however, are unlikely to have occurred for as long as our ancestors remained nomadic and did not settle in large communities. This changed with the adoption of a settled way of life and the development of animal husbandry. About 10,000 years ago in the Middle East things had advanced to the point at which people abandoned their nomadic way of life in favor of a more comfortable and settled existence growing crops and rearing animals. Within a few thousand years this altered way of life spread as far as Central Europe. And it was indeed a radical change in that human beings had made themselves independent of nature and instead had begun to subordinate nature to their interests.

People were now living closer together. Settlements evolved into towns. The hour of the epidemic had definitively arrived. At the time of Christ there were probably only a few hundred million people on Earth, then by about AD 1800 there were one billion. By 1950 the world's population had passed 2.5 billion, and by 1975 four billion. And then in October 1999 the United Nations symbolically welcomed a baby born in Sarajevo as the world's six billionth inhabitant. Since then the world's population has reached about 6.5 billion. And sometime between 2005 and 2010 another milestone will have been passed: by that time every second person in the world will live in a city. With this population density and with increasing globalization, the potential for epidemics is clearly greater than ever.

Many scientists now refer to three great waves of disease in human history. Domestication of animals and the growing of crops laid the basis for the first epidemics. These probably occurred about 10,000 to 20,000 years ago. Close proximity between humans and domestic animals led to a marked increase in zoonoses, while close contact between human beings permitted transmission of disease by droplet infection, direct skin contact, fecal-oral contact, and contamination of drinking water with fecal microorganisms. At the same time, farming and animal husbandry made life easier for mosquitoes, flies, fleas, ticks, and midges and consequently favored the transmission of diseases such as malaria, yellow fever, and dengue by these vectors. These classical infectious diseases have gradually become less common in industrialized countries, but remain rife in developing countries.

At present we are in the middle of a second wave of disease. The characteristic feature of this is that most present-day epidemics are largely independent of pathogens. Diseases such as cancer, cardiovascular disease, diabetes, obesity, and chronic inflammatory diseases are the most prominent such diseases. They have become increasingly common since the start of industrialization. These new epidemics have arisen firstly because of our success in the fight against classical infectious diseases, and secondly because of changes in lifestyle and increasing life expectancy in industrialized countries.

And now a third great wave of disease can be seen on the horizon. Like the first, this will be directly related to pathogenic organisms. Industrialization of animal husbandry and of food processing, increasing proximity between civilization and virgin forest, and globalization and increasing migration are preparing the ground for new pandemics. A harbinger of this was the pandemic of Spanish influenza that occurred at the end of the First World War, while HIV/AIDS shows that this change has now occurred. The next threat – possibly in the form of an avian influenza virus of the aggressive strain H5N1 that becomes pathogenic to humans – could be just around the corner.

4.2 Microbes: ancient jacks of all trades

A giant leap for mankind, a small step for microorganisms: in the time-scale of microorganisms the 20,000 years of their coexistence with human beings are no more than a moment. Their world began more than three billion years ago. When human beings first walked upon the Earth, microorganisms had already occupied every available niche, including that provided by our ancestors. Bacteria live with equal success on mountain peaks and in the depths of the oceans, in virgin forests and in cities, in the coldest glacial lakes and in the hottest geysers. Even on the deepest seabeds, where hot springs at temperatures of up to 250 degrees Celsius bubble up from the Earth's interior, certain species of bacteria are to be found.

I don't know whether we are able to determine the mass of all the bacteria in the world with any degree of accuracy. I am sure, however, that Tom Gold is correct in his calculation that bacteria contribute at least as much biomass to the world as all the plants and animals put together. Microorganisms have established themselves in every possible niche. Only a small proportion have chosen humans, animals, or plants as their home.

What is the secret of microorganisms? In principle it is the speed with which they reproduce and thus the rapid rate at which they can change. Many bacteria go through one complete replication cycle in half an hour. Evolutionary changes comparable to those that human beings, with their long intergenerational time, take thousands of years to achieve take bacteria only a few days to accomplish. Even if most of the changes that occur prove to be harmful or even fatal to the newly formed microorganisms, the few variations that prove to be beneficial soon get the upper hand and conquer new biological niches. This is "Darwin at his best".

Human development has taken a different path, namely that of division of labor by means of specialization and complexity. We like to see ourselves as the most complex of all living creatures and as the "crown of creation". The various human organs perform different tasks in the body. The stomach, intestine, and liver have the task of digesting food, the lungs that of breathing. The heart drives the blood circulation, and the brain is responsible for perception and thought. Responsibility for combating pathogens is borne by the immune system. Thanks to our intellectual ability, we humans have also been able to develop aids such as antibiotics and vaccines. Speed and ability to change or specialization – only time will tell which of these strategies will prove more successful in the long term.

4.3 Cooperation, coexistence, conflict

The human intestine is one of the most densely populated regions imaginable: in rough terms, the intestinal canal of each and every one of us contains about 1000 times more bacteria than there are human beings on the face of the Earth. A mass of 90 grams of bacteria – the approximate weight of the bacterial flora of our intestine – contains the scarcely believable number of one trillion (10^{12}) to one hundred trillion (10^{14}) bacteria. Yet there is no evidence of any overpopulation. On the contrary, it is only the presence of the intestinal flora that permits proper digestion. This coexistence of humans and microorganisms creates a whole range of possibilities. Our attention is mostly drawn to illnesses, however they are only one side of the coin.

So far no pathogen has succeeded in wiping out mankind. Nevertheless, there have been some devastating epidemics. In the Middle Ages, for example, the plague carried off between a quarter and a third of the population of Europe. Shortly after the end of the First World War, Spanish Influenza claimed the lives of about 50 million people in a period of only two years. Influenza thus probably holds the record for the disease that has caused the greatest number of deaths within a given period of time. Overall, however, i.e. independently of time, the disease that has caused the greatest number of deaths is probably tuberculosis. Yet now AIDS is catching up at an alarming rate.

Conversely, attempts by humans to wipe pathogens off the face of the Earth have met with precious little success to date. The only proven example of complete elimination of a pathogen is that of the smallpox virus. In 1980 this was officially declared to have been eradicated. Nevertheless, smallpox viruses persist in at least two laboratories, namely in Atlanta, USA and in Novosibirsk, Russia.

From disinterest to dependency

The range of interplay between microbes and mankind is extraordinarily broad. Most microorganisms have absolutely no interest in human beings. Only a disappearingly small proportion even come into contact with human beings; of these, a few live inside our bodies.

Let's start with the pleasant part, namely those fellow occupants of our body from which we benefit. These are known as symbionts, meaning microorganisms that benefit us but also derive benefit from us. Over the course of millions of years we have become so accustomed to a few of these that life without them would be quite unimaginable. Mitochondria, the power stations of cells in which the body produces energy in the form of chemical compounds, are an example of this. At some point in the course of evolution cells allowed bacteria to enter them on the basis of a "pact" to the effect that the bacteria would obtain nutrients from the cell and in return would release energy. Step by step the bacteria then underwent a process of regression until eventually they were able to perform one task and one task only, namely to produce energy. Nowadays our entire metabolism depends on the presence of mitochondria. For their part, they are no longer able to live alone, though they still have their own genetic material. As a humorous sideline, it could be said that the presence of this mitochondrial DNA provides the answer to the eternal dispute between mother and father as to which of them has passed on more of their genes to their child. Answer: the mother. This is because mitochondrial DNA is passed on only via the ovum. Spermatozoa, by contrast, have transferred the power supply for their motility into their tail, which they cast off on fusing with an ovum. It must be admitted, however, that mitochondrial DNA does not transmit any unique characteristics to the next generation.

A similar form of symbiosis evolved also in plants: chloroplasts, which trap solar energy and generate oxygen in a process known as photosynthesis, are likewise descended ultimately from bacteria.

In a high proportion of cases the reasons for the coexistence of microorganisms in our body are still a mystery: we do not know whether the microorganisms concerned are beneficial to us or of no significance whatsoever. Such microbes are known as *commensals*. Among other things, symbionts and commensals form the normal flora of our intestine. We know that some of these assist in the process of digestion, whereas in the case of many others we know too little about them to be able to describe them as being useful.

Benign invasion

Containing as it does about 90 grams of microorganisms, the intestine is the part of the human body that is most densely populated with microorganisms. More than 400 different bacterial species have been identified to date, and the total number of species present is believed to be about 500. As early as the moment of birth microbes pass from the maternal vagina to the newborn infant. In a few cases some of the mother's intestinal bacteria even find their way into the infant's intestine at this time. Within a few days of the infant's birth this process of colonization is largely complete. The first to establish themselves are aerobic organisms, that is to say organisms which, like human beings, require oxygen in order to survive. When these organisms have used up all the available oxygen, in other words when anaerobic conditions have been created, colonization by anaerobic bacteria, i.e. bacteria that do not require oxygen in order to survive, starts. And even at this early stage antibiotic-resistant bacteria can start colonizing the intestine.

The normal intestinal flora benefits us in all sorts of ways. By occupying the docking sites on the cells of the intestinal wall, bacteria prevent colonization by foreign intruders. At the same time they maintain a milieu that is typical of the intestine, produce substances that actively inhibit opponents, and consume nutrients.

In addition, the normal intestinal flora trains the immune system and keeps it prepared at all times for any attack by pathogens. For this reason mice that are reared in a completely microorganism-free environment have a markedly impaired immune system.

Not only do we benefit from this protective effect, but our ability to digest food depends at least in part on the presence of microorganisms: the human enzymes that are present in the stomach and intestine are unable to break down many complex sugars and some proteins, while other proteins are not easily absorbed. In order to break down and utilize such nutrients, the body requires the assistance of the normal intestinal flora. Absorption of these nutrients occurs in the large intestine, the "principal realm" of the intestinal bacteria. The intestinal flora also produces vitamins, that is to say trace substances that we need but are unable to produce ourselves. These include vitamin K, folic acid, vitamin B_{12}, vitamin D, biotin, nicotinic acid, and thiamine.

In addition, the correct intestinal flora renders certain toxins harmless. Many toxins, especially those formed from amino acids when meat is fried or grilled at high temperatures, are suspected of favoring the development of cancer. A "good" intestinal flora breaks these substances down.

The intestinal flora can be favorably influenced by probiotics and prebiotics. *Probiotics* are living organisms such as bifidobacteria and lactobacilli that favorably influence the intestinal flora. *Prebiotics* are nutrients that indirectly influence the composition of the intestinal flora; they include indigestible sugars such as inulin and oligofructose. They are resistant to breakdown in the upper gastrointestinal tract and can be digested only by certain useful intestinal bacteria that are found in the large intestine. For example, the complex sugar inulin, unlike its component fructose, increases the proportion of bifidobacteria in the intestine, since these bacteria are able to digest this sugar. Probiotic and prebiotic diets are most effective when taken together, such a combination being referred to as a symbiotic diet. Probiotics and prebiotics are meeting with an increasingly enthusiastic reception, especially among manufacturers of so-called *functional food*. Foods containing additives that promise better health are very much in vogue at present – and are having an enormous economic impact. Nevertheless, hard scientific proof of their effectiveness is often in short supply. For example, the claim that the consumption of yogurt, as well as having a positive effect on the intestinal flora, has more general benefits for health is as yet unproven.

Gentle resistance and aggressive attack

At the other end of the scale are microbes that harm or cause disease in their host, i.e. in human beings, as a matter of course. Here again there is a broad range. Some pathogens cause illness whenever they infect a human being. These include poxviruses and HIV, though there may be the occasional human being who can carry HIV for a long time without becoming ill. Other microorganisms cause illness only very rarely and only when they find themselves in the wrong place or when the host's immune system is compromised. Such microbes are known as *opportunistic* pathogens. They include the normal inhabitants of the mouth and pharynx, some vaginal microorganisms, and many intestinal microorganisms.

Even organisms that cause life-threatening diseases such as tuberculosis do not always cause illness. Thus, 90% of people infected with the causative agent of tuberculosis do not become ill, but carry the organism without being aware of it. In principle, therefore, even this bacterium, which has caused more human deaths than any other microorganism, can be regarded as an opportunistic pathogen. The task of keeping the agent of tuberculosis under control is undertaken by our immune system. If this is weakened, the disease breaks through, i.e. active tuberculosis develops.

It is often argued – and in most cases it is true – that in the course of the common evolution of microbes and humans the microbes become less harmful. This makes evolutionary sense, since a microorganism is more likely to spread if it doesn't kill its host immediately, but instead makes use of its host for as long as possible and allows itself to be spread by the host. Organisms that cause illness are known as *pathogens*. These live at the expense of the host and in the normal course of events cause harm that we experience as illness. Pathogens are thus also *parasites*. The ability of an organism to cause disease is known as the *pathogenicity* of that organism. Similarly, the mechanism by which a disease develops is known as the *pathogenesis* of that disease. Another commonly used term is *virulence*. This refers to the degree of aggressiveness of a pathogen and thus to the severity of the illness that it causes. Accordingly, highly virulent organisms are dangerous, whereas avirulent organisms have more or less lost their pathogenic properties. Techniques of *attenuation* are used to artificially reduce the virulence of microorganisms. This is the critical step in the development of a live vaccine from an originally dangerous, i.e. virulent, pathogen.

The penetration of a pathogen into its host marks the beginning of *infection*, i.e. a conflict between microorganism and host (Fig. 6). This goes on for a certain amount of time without resulting in any apparent illness. This is referred to as latent infection. With certain pathogens this phase can last for a very long time. In some cases illness never develops. If the immune system fails to kill the pathogen, the pathogen persists in the body and the

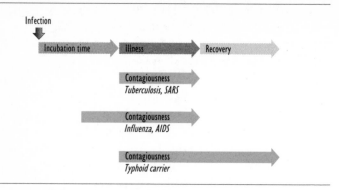

Figure 6   From infection through illness to recovery

The penetration of a pathogen into the body and the ensuing conflict with the host are known as infection. This begins with an incubation period during which the pathogen reproduces without causing symptoms of illness. Some infections are contagious at this stage. After the incubation period comes illness, during which the patient suffers specific symptoms. Generally, the diseased patient is contagious. The severity of disease depends upon the pathogen and the host. Illness is followed by recovery. In some cases, e.g. typhoid, pathogens can persist in niches within the body without causing illness. As they are nevertheless excreted, their host, known as a chronic carrier, remains contagious.

host becomes a silent carrier. Many people who are infected with the tuberculosis bacillus fall into this category before becoming ill, if at all.

In other cases symptoms appear very soon after the infection begins. Many microorganisms that cause diarrhea produce symptoms within hours. In the case of most pathogens symptoms appear after a few days to weeks. The phase between the beginning of infection and the onset of illness is known as the *incubation period* (Fig. 7). During this period some pathogens are screened off to such an extent that they cannot escape from the body and therefore cannot infect anyone else. This is the case with tuberculosis, for example. HIV, by contrast, can be transmitted before the clinical picture of AIDS develops. Influenza

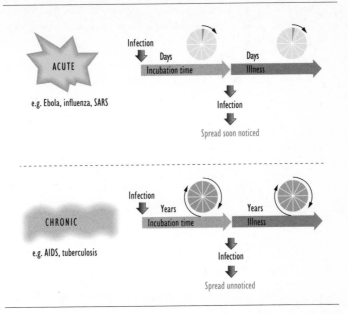

Figure 7   Principles of acute and chronic infections

Acute infections last days to weeks. A short incubation period of hours to days is followed by an illness that lasts days to weeks. Fatal acute infections such as Ebola, Spanish influenza, and SARS spread with great rapidity but are generally detected at an early stage. Chronic infections have an insidious course. After infection the pathogen can remain dormant in the host for years. If such a pathogen is infectious, as in the case of HIV/AIDS, it can spread unnoticed. In most cases the long incubation period is followed by a protracted illness during which the patient remains infectious. Unnoticed spread during the long incubation period enabled HIV/AIDS to spread throughout the world.

viruses are likewise passed on to other individuals before illness develops. Conversely, one of the reasons why the outbreak of the lung disease SARS was so limited was that the virus became infectious only after the onset of illness. The way in which a pathogen behaves during its incubation period can thus be a critical determinant of the extent to which an infectious disease spreads.

The recovery phase of illness can likewise take different forms. In the most favorable case the immune system has completely eliminated the pathogen. In other cases a few microorganisms persist in the now healthy person.

Doctors are required to report such chronic carriers to the health authorities. The persons concerned are then kept under surveillance and treated until the pathogen has been completely eliminated. Until this has been confirmed they are prohibited from working in occupations that can bring them into close contact with other people.

Many pathogens have a very narrow range of hosts. The causative agent of syphilis, for example, affects only human beings. Other pathogens can make animals and humans equally ill and can jump back and forth between species. Such diseases are known as *zoonoses*. The influenza virus, for example, is at home not only in human beings but also in pigs and many species of poultry. Ebola viruses kill chimpanzees, gorillas, and human beings with equal effectiveness. Three-quarters of all newly appearing infectious diseases are zoonoses.

The doctor and author Frank Ryan has put forward the following hypothesis about the coexistence between human beings and microbes, which he describes as an "aggressive symbiosis": In the jungles of Borneo certain ants have developed a special way of coexisting with plants. They drink the sweet sap of the plants and in return for this privilege attack plant-eating animals in order to safeguard the source of their food. A similar mechanism can be postulated for certain infectious diseases. Imagine a microorganism that thrives in a host and does the host no harm. Any animal that tries to eat the host is made ill and killed by the microorganism – in a sense as a form of revenge or punishment for an action that harms not only the intended victim but also the microorganism. The intention of the microorganism is to protect its host in order to preserve its own biological environment. According to this hypothesis of aggressive symbiosis, Ebola viruses leave their (as yet unidentified) primary host unharmed in order to use it as a place to live, whereas they are lethal to primates that have destroyed this primary host. HIV is similar in that it is essentially harmless to the nonhuman primates in which it originated, whereas it kills humans. However, whether this "punishment" really exerts selective pressure in the Darwinian sense after the damage has been inflicted remains a matter of speculation.

Fragile balance over thousands of years

In view of all this complexity it is not surprising that different microorganisms have developed different strategies for survival. The same is true on the host side of the relationship. As Darwin explained, those individuals who are better protected against a pathogen tend to be selected. These mechanisms have played a major role in human evolution.

Malaria is an instructive example of this. Malaria plasmodia, the single-celled organisms (protozoa) that are responsible for malaria, live in red blood cells and use the blood pigment hemoglobin as a source of food. People who suffer from hereditary sickle-cell anemia have an altered form of hemoglobin. Malaria pathogens are unable to use this form of hemoglobin in the usual way. People who suffer from sickle-cell anemia are therefore partially protected against malaria. Sickle-cell anemia carriers are substantially less likely to die from malaria than are unaffected people. In this way the disease has contributed to the spread of at least one form of sickle-cell anemia in West Africa. Whereas the percentage of the world's population that carries the trait for sickle-cell anemia is around 5%, in the West African countries of Ghana and Nigeria around 15 and 30%, respectively, of the population carry the trait, while in Uganda up to 50% of certain tribes are affected.

In infectious diseases such as tuberculosis in which the body succeeds in controlling the pathogen before active disease occurs, both parties, i.e. pathogen and human, are likewise subject to strong selective pressure. As recently as the beginning of the 19th century tuberculosis was the leading cause of death in European cities. At the same time resistant individuals were being selected in Europe. By contrast, Eskimos had no contact with tuberculosis until the middle of the 20th century – and consequently are now far more susceptible to this disease.

Strategies of microbes

Togetherness leads to interdependence. Over millions of years microbes have developed various ways of ensuring their own survival. As with the immune system, use of military metaphors can help illustrate strategies of attack. Humans have used such strategies only in large-scale contexts – from chemical and biological warfare to trench warfare.

Some pathogens don't harm humans directly. As mentioned previously, some pathogens form toxins and practice a kind of chemical and biological warfare. This group includes the causative agents of diphtheria, anthrax, and tetanus. Strictly speaking, these pathogens are not required at all for the illness to develop: the toxin is enough. Vaccines directed against toxins have been available for a long time. Here again, the antibodies do not attack the pathogen itself, but instead eliminate the toxins.

A blitzkrieg approach is also adopted: certain viruses and bacteria attack the host rapidly and very directly. Such pathogens include Ebola viruses, influenza viruses, common cold viruses, other pathogens that cause respiratory diseases, and pathogens that cause diarrheal diseases. This is by far the biggest category of pathogens. Considered as a group, it causes about eight million deaths per year – more than tuberculosis, HIV/AIDS, and malaria combined, which between them cause up to six million deaths per year.

From the point of view of the pathogen, of course, the symptoms are often very useful for spreading the disease – a runny nose and diarrhea, for example, are absolutely ideal in this regard. Sneezing propels viruses out into the environment at a phenomenal speed. Coughing, spitting, and nose blowing are ideal from the point of view of pathogens. Diarrhea releases huge numbers of microbes which can then find their way into new hosts very rapidly by direct fecal-oral transmission or via contaminated drinking water. Ebola viruses and the agents responsible for similar diseases damage blood vessels and thereby cause bleeding that releases them into the external environment.

All these strategies are extraordinarily effective. The blitzkrieg approach takes the immune system by surprise, but in subsequent infections the immune system is easily able to recognize and attack the pathogens. Vaccines are therefore useful against many pathogens that adopt the blitzkrieg strategy. However, the pathogens then take countermeasures to defend themselves against the vaccines. Influenza viruses, for example, change with lightning speed, constantly develop new forms of camouflage, and conceal themselves from the immune system. Adequate protection can therefore be provided only by newly composed vaccines that are modified on an annual basis.

Some pathogens have developed a sort of trench warfare against which antibodies are useless. Such pathogens hide themselves away in host cells, well protected from attack by the immune system, and live in a state of apparently peaceful coexistence with the infected host. In many cases they cause no harm to the host cell at first. Some such pathogens, for example herpesviruses, can remain in their hiding places out of sight of the immune system for years until the disease breaks through in response to stress of some kind. Herpes-induced cold sores of the lips are a well-known and unloved example of this strategy. Most hepatitis viruses likewise persist over long periods; some do not impair liver function, whereas others can cause cirrhosis, or even cancer, of the liver after many years.

Mycobacterium tuberculosis also fights by trench warfare. This robust microorganism survives inside human immune cells that manage to keep it in check but are unable to kill it. Inside the host cell the pathogen is protected from antibodies. T lymphocytes, however, are able to locate infected cells. T killer cells kill captured cells directly, while T helper cells mobilize defense mechanisms inside the infected cells.

Another group of pathogens take this approach one step further by waging guerilla warfare in the sense that they deliberately infiltrate the immune system. HIV, for example, knocks out the headquarters of the immune system. It enters T helper cells via their CD4 markers and eventually eliminates these immune cells completely, thereby causing the host's immune system to collapse. Typically, therefore, AIDS patients die not from the HIV infection itself, but from secondary infections such as tuberculosis and fungal diseases. In addition, by constantly changing its appearance HIV is able to permanently avoid the recognition mechanisms of the immune system. This constant changing of appearance is a major obstacle to the development of an effective vaccine against HIV.

A similar strategy is that of fomenting a civil war. Pathogens that act in this way trigger autoaggressive reactions by fooling the immune system into fighting against the body's own cells. *Trypanosoma cruzi*, the parasite that causes Chagas' disease, belongs to this group. This chronic disease causes progressive heart failure, among other things. According to *Médecins Sans Frontières*, about 18 million people, mostly in South America, are infected with this pathogen.

Other more or less dangerous microbes probably also belong to the group that wages guerilla warfare. We know that many autoimmune diseases are either triggered or promoted by infections. An example that was previously common, but is now rare, in industrialized countries is that of rheumatic fever occurring in the aftermath of a streptococcal infection. This disease is based on a similarity between certain surface structures of streptococci and heart muscle cells. The antibodies formed in response to the infection attack not only the pathogens but also heart muscle cells, thereby causing inflammation in these.

4.4 Pandemic, epidemic, or what, precisely? The technical jargon used by epidemiologists

"According to a recent study, German hospitals do not have enough intensive care beds to cope with a possible severe epidemic of influenza. Experts consider that human influenza viruses or strains of avian influenza virus that are dangerous to humans could cause such an epidemic. As reported in the Hamburg-based weekly newspaper *Die Zeit*, bed shortages could occur especially in cities. In a moderately severe epidemic with a morbidity rate of 30% of the population, the only German state with a sufficient number of intensive care beds would be Saxony-Anhalt, according to the study, which was conducted by Allianz Insurance (Munich) and the Rhenish-Westphalian Institute for Economic Research (RWI Essen). Depending on the severity of such a pandemic, Germany's gross domestic product would shrink by between 1.0 and 3.6%."

This is the first part of a report such as might appear at any time in any newspaper in the industrialized world. This particular report was published in July 2006. Its choice of vocabulary is striking: epidemic, pandemic, morbidity rate. These specialized medical terms have moved out of hospitals and specialist circles. However, whether they are correctly understood by all is quite another matter. We talk about epidemics and pandemics all the time, but do we know what these terms actually mean? Let's therefore take a closer look at a number of terms that are central to the topic of infectious diseases. The press report quoted above uses specialized terms from the field of epidemiology, the science that deals with the spread of diseases including infectious diseases. As we have seen from many examples, the extent to which a disease spreads depends upon the pathogen, the host, and environmental factors.

As we have already referred to *pathogen factors*, we will list only the most important factors here. The host range of a pathogen indicates the range of organisms that a pathogen can endanger. Is the host range small, as in the case of the feline immunodeficiency virus, which affects felines but not humans? Or is it broad, as in the case of salmonellae, which can simultaneously infect poultry, humans, and dogs? Other pathogen factors include pathogenic potency, i.e. virulence, and infectiousness, i.e. the potential to infect other hosts. Also worthy of mention are factors that enable the pathogen to survive in humans and animals (including intermediate carriers, i.e. vectors) and in the environment. The latter type of factor includes resistance to soap, detergents, and disinfectants, resistance to attack by the immune system (whether this be triggered by the infection itself or by vaccination), and of course resistance to antibiotics.

Host factors are factors related to the carrier of the pathogen. These include genetically determined resistance, natural resistance, and immune mechanisms. Increasingly important in this regard, however, are other factors such as cultural and social circumstances and modes of behavior such as poverty, nutritional state, travel, migration, sexual behavior, hygienic behavior, and contact with other people.

Environmental factors are factors that influence the survival of pathogens in the environment. They include climatic conditions such as temperature and precipitation as well as the use of antibiotics and pesticides in farming. Environmental factors are also important for vectors. There is considerable discussion at present about the influence of climate change, i.e. global warming, on many insects. For example, the geographic range of certain species of midges and ticks could spread. In 2006 bluetongue, an African disease of animals that is transmitted by certain midges, was found in Germany for the first time. In 2002 there was an outbreak of West Nile fever, a disease that is dangerous to humans, in the USA. This disease, which is transmitted by midges from a reservoir in birds, originated in Africa and the Middle East. Climate change may also increase the geographic spread of the malaria mosquito. In the 1950s and 1960s the insecticide DDT was used to combat the malaria mosquito. Because of serious concerns about the environmental damage that DDT can cause, there was then a period of twenty years during which use of this insecticide was not recommended. In 2006 the World Health Organization did an about-face, once again recommending targeted indoor use of the insecticide to fight malaria (see also box: "DDT and malaria control", page 128). Equally important measures are the draining of swamps, correct waste disposal, and measures to combat environmental pollution in towns. What is clear is that many environmental factors are also host factors in the broadest sense, since they are brought about – whether for good or for bad – by human beings.

As an example of this, cigarette smoke damages the lungs and is therefore a risk factor for infectious diseases of the airways (respiratory tract). From the perspective of the (active) smoker it is a host factor, whereas from that of the passive smoker it is an environmental factor.

Infectious diseases are not just infectious diseases. In order to describe the severity of an infectious disease more precisely, epidemiologists use three terms, at least two of which crop up again and again in reports about disease.

An *epidemic* (from Greek *epi*, upon + *demos*, people) is a spatially and temporally limited outbreak of a disease, e.g. plague in the Middle Ages.

An *endemic* (from Greek *en*, within + *demos*, people) is the occurrence of a disease in spatially limited but temporally prolonged fashion, e.g. smallpox, which remained prevalent in some parts of the world until the 1970s.

A *pandemic* (from Greek *pan*, all + *demos*, people) is a temporally limited but worldwide outbreak of a disease. The outstanding example of a pandemic in recent times is HIV/AIDS, however the term has been used more frequently in relation to a feared pandemic of avian influenza due to H_5N_1 viruses. Spanish influenza was also a pandemic. This will be discussed in more detail later.

Even brief, local outbreaks of an infectious disease can show different patterns:

- Many chains of infection originate at a certain time from a certain person, known as the *index case*. The shorter the incubation period, i.e. the period between infection of an individual and the onset of illness in that individual, the more rapidly epidemiologists are able to work out the sequence of infection, since in such cases the affected individual has symptoms. An example of this is the outbreak of SARS that was traced to a single infected guest of the Metropole Hotel in Hong Kong (see Section 5.9).

– There are also chains of infection that originate from a single source but persist over a longer period. An example of this is diarrhea caused by contaminated foods in residential institutions. The source of the infection in such cases is often egg products contaminated with salmonellae or, for example, a cook with a suppurating wound who mixes salads by hand. In the case of an outbreak of hepatitis A that occurred among tourists in an Egyptian hotel, the chain originated from an infected employee who squeezed fresh orange juice. Similarly, many outbreaks of hepatitis B have been attributed to the use of contaminated tattooing equipment.

– In many cases diseases spread like an avalanche from a number of people via a number of chains. This is how children infect each other with measles in schools – far too often, since such transmission is avoidable. In the fall of 2006 statisticians identified over 2200 such cases in an outbreak that occurred in North Rhine-Westphalia.

For the sake of completeness, we should explain the meaning of a few more terms:

Morbidity is the proportion of ill people in a population, expressed mostly per 100,000 people. The proportion of people affected by a disease who die from that disease is the *mortality* of the disease, likewise expressed mostly as deaths per 100,000 people. The *incidence* of a disease is the number of new cases of that disease, expressed mostly per 100,000 people, occurring within a given period, typically a calendar year. And finally, the *prevalence* of a disease is the number of people affected by the disease at a given point in time, likewise expressed mostly per 100,000 people.

The World Health Organization expresses data on diseases and the socioeconomic burdens imposed by diseases mostly in terms of three parameters: death, years of life lost, and disability-adjusted life years. More informative than the pure number of deaths is the parameter "years of life lost" (YLLs), meaning the number of years of life that people lose by dying prematurely. With this parameter death in the early years of life is weighted more heavily than death at an advanced age. The most informative parameter, however, is "disability-adjusted life years" (DALYs), which consequently will be used frequently in the following discussions. This parameter indicates the number of healthy years of life lost to disability or illness. The DALY is thus primarily a measure of health, however it also permits direct conclusions as to loss of productivity in a society affected by illnesses. Like the YLL, the DALY takes age into account. The first few years of life and the years of life spent at an advanced age are given a lower weighting than the economically productive years of life. Finally, account is also taken of the severity of illness on a scale extending to death. The DALY is thus a much better measure of the economic and social impact of a disease than are parameters such as mortality and morbidity.

4.5 Undesirable alliances: how pathogens play a role in cancer and other diseases

Cancer

Every year 11 million people worldwide are diagnosed with cancer. According to the World Health Organization, almost eight million people, more than two-thirds of whom were from poorer countries, died of malignant tumors and their complications in 2007. Cancer was thus responsible for 13% of all deaths.

However, a little-known fact is that every fifth case of cancer in humans is attributable to microorganisms. And the true figure could be higher than this, since only 10% of all forms of cancer are known for certain to have nothing to do with microorganisms, whereas in the remaining 70% it is entirely possible that microorganisms may play a role as carcinogens or co-carcinogens.

The best-known microbial carcinogens are the bacterium *Helicobacter pylori*, which causes peptic ulcers and certain forms of stomach cancer, human papillomavirus (HPV), which causes cervical cancer in women, and hepatitis B and C viruses, which can cause liver cancer. However, there are many more.

The precise mechanisms by which viruses, bacteria, and parasites cause cells to degenerate are highly varied and have yet to be fully elucidated. Viruses, for example, use cells as a platform for replication. To do this they turn the cell factory upside down and cause it to mass-produce viral proteins. In most cases the affected cell eventually dies and thereby releases large amounts of virus. In some cases, however, the cell does not die. If, at the same time, other cellular control mechanisms have been put out of action, the cell may degenerate. The first step on the road to cancer has then been taken, since in principle a single degenerate cell can grow to form a tumor. At the same time a vicious circle has been created in that the degenerate cell is also more susceptible to harmful external influences. Ideally, the immune system will identify the troublemaker and eliminate it before it can reproduce. In ill people, however, this balance has been lost. In AIDS, for example, the immune system is a mere shadow of its former self. This is one reason why many AIDS patients are also afflicted by cancer.

Helicobacter pylori, stomach ulcers, and stomach cancer

Among bacteria, *Helicobacter pylori* has achieved fame as a cause of cancer. It was the first bacterium to be deemed a bacterial carcinogen by the World Health Organization. The discovery of this microorganism in the 1980s was momentous in that it obliged the medical profession to reclassify stomach ulcers – which until then had been held to be a consequence of stress, nervousness, suppressed anger, or incorrect diet – as an infectious disease. Whereas up to then treatment had been symptomatic in that it was aimed at neutralizing the enormous amounts of gastric acid that were produced, antibiotics were now prescribed. This meant that doctors were now often able to heal stomach ulcers – and in so doing prevent stomach cancer. This, however, is not always enough: in some people gastric acid production remains high even after the pathogen has been eradicated. Alcohol and aspirin can also cause gastritis.

Around three billion people, i.e. almost half the world's population, are probably infected with *H. pylori*. According to the World Health Organization up to 70% of people in developing countries and between one in five and one in three people in industrialized countries are infected. More than one million people in the UK and almost 400,000 in Australia suffer from stomach ulcers. Around 25 million people in the USA and four to six million Germans suffer from stomach ulcers at some time in their life. Approximately 1% of infected people develop stomach cancer. *H. pylori*-infected people are six times more likely to develop stomach cancer than are people whose stomachs do not contain this bacterium. In the USA almost 12,000 people die of stomach cancer every year, while the corresponding figure for Germany is about 8000. At the beginning of the 20th century malignant tumors of the stomach were among the most common types of cancer in industrialized countries. Since then the rate of stomach cancer has fallen considerably in the industrialized world – thanks to improved hygiene and the use of antibiotics,

A success story, then? That's how it seemed at first, however the reality may be more complex. There is increasing evidence that mass use of antibiotics and improved hygiene in Western countries may replace one evil with another.

In parallel with the reduction in the number of cases of stomach cancer there has been a steady increase in the incidence of esophageal cancer, a particularly aggressive form of cancer that used to be almost unknown. In the USA, for example, the incidence of this form of cancer is increasing by 10% per year. And esophageal cancer is one of the deadliest forms of cancer, with many victims dying within a few years.

How is this increase to be explained? Epidemiologic studies have shown that the risk of developing esophageal cancer increases after successful eradication of *H. pylori*. It turns out that as well as causing gastric ulcers, the bacterium reduces peak production of gastric acid, which, by flowing back into the esophagus, is principally responsible for degeneration of esophageal cells. And it is precisely those bacterial strains that most commonly cause cancer that do this best. They set upon the gastric mucosal cells with a special injection apparatus present on their surface while at the same time indirectly protecting the mucosa of the esophagus. Thus, the intervention by doctors may have had the effect of meddling with a complex interaction between pathogen and host that has developed over millions of years of evolution.

In order to avoid any misunderstandings, let's go through the sequence of events once again: *H. pylori* initiates an inflammatory reaction in the gastric mucosa that can lead to the development of stomach cancer. Prompt eradication of these microorganisms prevents the inflammatory reaction and the development of stomach cancer. At the same time, however, eradication of *H.*

pylori leads to damage to the esophagus due to reflux of gastric acid. This in turn can lead to the development of esophageal cancer. However, since reflux of gastric acid into the esophagus can be blocked by drugs, the potential vicious circle between *H. pylori* eradication and esophageal cancer can be broken.

Viral carcinogens

Human papillomavirus, HPV: This is another microbial carcinogen that has made headlines, particularly in 2006, when a vaccine directed against certain strains of papillomavirus was licensed in many countries. According to proponents of the vaccine, a course of three injections could prevent countless cases of cancer and cancer-related deaths. This is because the viruses can cause cancer of the cervix. In the USA almost 4000 women die from cervical cancer each year, while in Germany 6000 to 6500 women develop the disease each year and 1800 die from it. Each year 1.5 million women in Australia, 4.5 million women in the UK, and 5.5 million women in the USA become infected with HPV. However, the virus is a far bigger problem in developing countries: worldwide, about half a million women suffer from cervical cancer and half of these die as a result of it.

HPV is transmitted by sexual contact, and not infrequently. More than two-thirds of all women become infected at least once in their lifetime with at least one strain of this pathogen. In the USA 20 million men and women are carriers of HPV. In most cases the virus goes unnoticed and is destroyed by the immune system. In some women, however, the virus hides in the body, and some of these women develop genital warts. Men can also be affected. A small proportion develop cancer. Years, and sometimes decades, pass between infection and the appearance of cancer. HPV is the most common cause of cervical cancer, which in turn is the second most common form of cancer in women.

The new vaccine could prevent up to 70% of deaths due to cervical cancer, since it acts against the viruses that are responsible for about 70% of cases of this type of cancer. However, it only stimulates production of antibodies that prevent the virus from colonizing the cervical mucosa in the first place. Once the viruses have entered cells, the vaccine is ineffective. The message is clear: girls need to be immunized at a very early age, i.e. before their first sexual contact.

Women who are already infected need T cells, or to be more precise T killer cells. Work is now under way to develop a new generation of HPV vaccines capable of destroying infected cells with their resident HPV. Such vaccines would be suitable for post-exposure eradication.

Hepatitis viruses: All hepatitis viruses have one thing in common: they cause liver disease. In principle, all of them can cause acute inflammation of the liver leading to jaundice, dark urine, nausea, vomiting, and abdominal pain. Nevertheless, hepatitis viruses differ greatly from one other, including in terms of their significance as pathogens.

Chronic liver diseases are responsible for about 1.5 million deaths per year worldwide; of these, about 660,000 are due to primary liver cancer (meaning that the tumor arose in the liver itself rather than being a metastasis of a tumor that arose somewhere else in the body). Almost 800,000 people suffer from chronic cirrhosis of the liver.

Hepatitis A viruses are mentioned here purely for the sake of completeness, since they do not cause cancer, but merely a form of acute hepatitis that does not become chronic and in most cases heals spontaneously. The viruses are excreted in the feces and the disease is often passed on via contaminated foods or by fecal-oral transmission. Hepatitis A is relatively easily acquired when traveling, e.g. by eating unwashed fruit and vegetables, unboiled water, or seafood. Not uncommonly the virus is spread by infected kitchen staff. Statisticians estimate that each year about 1.4 million people develop acute liver inflammation due to the hepatitis A virus. In 2006 the number of new infections with hepatitis A virus in the USA was estimated at more than 30,000. In the same year doctors reported about 1200 cases in Germany. A vaccine that protects against this disease is available and is recommended for people traveling to developing countries.

In global terms, *Hepatitis B* is one of the most important pathogens responsible for human hepatitis. Healthy hepatitis B carriers are the principal source of infection. As transmission occurs typically via blood or sexual contact, hepatitis B is found with particular frequency in drug addicts and homosexuals. Altogether about two billion people, i.e. a third of the world's population, have been infected with this virus at one time or another. About 350 million people have chronic infection. Most of these people have no symptoms, however every tenth patient with chronic hepatitis B will eventually develop cirrhosis and many will also develop liver cancer. The latter disease kills one million people per year.

The virus is highly prevalent in China and sub-Saharan Africa, whereas in Central Europe and the USA its prevalence is relatively low. An estimated 1.0 to 1.4 million people in the USA have chronic hepatitis B infection. In Germany 1200 cases were officially reported in 2006. As with hepatitis A, a vaccine that protects against hepatitis B is available. It is not exactly cheap, but is highly recommended for people at high risk.

An example of the protective effect that can be achieved by universal immunization in high-risk areas is provided by Taiwan.

This is one of the Asian countries in which hepatitis B infection is most common. In July 1984 the Taiwanese authorities made immunization against the virus compulsory for all children. Within a few years a dramatic reduction in the incidence of liver cancer in children aged between six and fourteen years became apparent: in the years between 1981 and 1986 approximately seven children per million developed liver cancer, whereas ten years later this figure had been almost halved.

Hepatitis C: The World Health Organization sometimes refers to this virus as a "viral time-bomb". About 3% of the world's population, i.e. about 200 million people, carry this virus, and in a high proportion of these people there is a risk that the infection may progress to cirrhosis or liver cancer. A fifth of all deaths due to chronic liver disease (almost 300,000 deaths per year) are due to hepatitis C. And there is no satisfactory form of treatment and no vaccine. What's more, no vaccine is even in sight, since the virus – like HIV – is constantly changing. Should hepatitis C continue to spread like in the past it may soon become a greater threat to humankind than HIV/AIDS.

The hepatitis C virus is transmitted via blood, therefore drug addicts are at greatly increased risk of acquiring it. Transmission via sexual contact is less frequent. In Germany there were about 7500 new cases of hepatitis C in 2006.

There are even more forms of hepatitis, known as hepatitis G, hepatitis H, hepatitis I, etc., however these will not be considered here. The important thing to know is that pathogens can also be carcinogens, either by a direct action or by acting as cofactors, and that immunization against such pathogens can also prevent cancer. Presently available vaccines stop the pathogens from spreading and infecting cells. In the longer term, however, it should also be possible to develop vaccines that destroy infected cells and the pathogens that have settled inside them. Such vaccines should therefore be of benefit also to infected people at increased risk for cancer and even to people who have already developed cancer.

Other diseases

There is a great deal of speculation at present as to whether microorganisms also play a role – either directly or as cofactors – in other chronic diseases that are not obviously due to an infection. Some researchers even regard the infectious diseases that have been identified to date as being no more than the tip of an iceberg. According to this view, microorganisms play an as yet unknown role in most diseases and it is our task to elucidate this role. In the case of certain autoimmune diseases this view is certainly correct. Also, researchers recently discovered the importance of the normal intestinal flora for obesity, a disease of increasing importance in developing countries and newly industrializing countries.

This shows that many diseases can no longer be regarded as unidimensional in the sense that a pathogen causes the disease by direct action. Instead, diseases are multidimensional systems in which a variety of microorganisms interact in complex ways with environmental factors and the genetically determined individual susceptibility of the human host. This will become an extraordinarily exciting field of research in the biology of infectious diseases. So far, however, this field is still highly speculative. For now, it is enough to state that the musings about a possible link between *Chlamydia pneumoniae* and cardiovascular disease and Alzheimer's disease, and about a possible role of the protozoon *Toxoplasma gondii* in schizophrenia, still have a long way to go.

5 More Than a Body Count: The Major Infectious Diseases

> An infected man could transmit the poison to others and spread
> the contagion to people and places merely by looking at them.
> Nobody found any means of resisting. Almost everybody who
> was in the east or in the regions further to the south and north
> therefore died abruptly after contracting the pestilence as if
> struck by a mortal arrow that caused an ulcer to form in his
> body.
>
> *Historia de Morbo*, Gabriele de Mussis (ca. 1280–ca. 1356),
> Lawyer from Piacenza, contemporary witness to the Black Death in Europe

In the television series 'Six Feet Under' people were always dying.
And so it is in real life. In 2005, for example, some 58 million
people on our planet died, according to UN figures. Compared
to the figures for living people and births, however, this figure
seems almost small: in the same year there were 6,514,751,000,
i.e. over 6.5 billion, people alive on Earth and around 136 million
babies born. Yet unless fleshed out with details, figures are no
more than the dry work of statisticians. Of the 58 million dead,
for example, around 11 million were children under five years
old, and a high proportion of these died of a small number of
diseases, many of which are infectious. These major infectious
diseases form the subject of this chapter. And if, in this chapter,
there is repeated reference to numbers of deaths and sick people,
this is not intended as a sort of morbid body count, but simply

as a means of illustrating the scale of the problem. Because what these figures ultimately show is the urgent need for action in healthcare systems worldwide.

5.1 From colds to pneumonia: respiratory infections are number one

Every year the usual wave of coughs, colds, and sore throats wends its way around the northern hemisphere. For a few days we feel worn out and run down, our bones ache, our head buzzes, and our nose is runny. And in most cases it's all over within a few days. That's the extent of our usual contact with viruses and bacteria of the human respiratory tract in this part of the world. Each year, on average, every second person suffers some sort of respiratory infection. No great problem, then? On the contrary. Because worldwide, four million people die from acute respiratory infections each year. This figure includes people who die from influenza, but not those who die from tuberculosis. As a chronic airway infection, tuberculosis is listed separately by the World Health Organization. The list of respiratory diseases is long. It extends from upper respiratory infections – colds and sore throats, as mentioned above – to infections of the middle ear and the paranasal sinuses. There are also diseases of the lower airways such as pneumonia (inflammation of the lungs). The pathogens responsible also form an enormous group. Colds and runny noses are due mostly to viruses, in particular rhinoviruses, influenza and parainfluenza viruses, and coronaviruses. The causative agents of whooping cough (pertussis), diphtheria, and bacterial pneumonia rank among the top ten bacterial causes of respiratory infections. For some of these diseases, notably diphtheria and whooping cough, effective vaccines are available. Until the 19th century diphtheria was the most important cause of the high childhood mortality that prevailed in many European countries, however later it was largely suppressed in that part of the world by vaccination. Recently, as a result of increasing complacency about immunization, there have been sporadic outbreaks of diphtheria in a number of regions of the world. Altogether, 17 million people still contract whooping cough each year and the disease claims 300,000 lives annually, with 90% of the deaths occurring in developing countries.

In addition to the causative agents of tuberculosis and influenza, to which separate sections have been devoted in this book, pneumococci also warrant special consideration. This form of streptococcus mostly infects children under two years of age and elderly people (over 65 years of age). In industrialized countries pneumococcal pneumonia mostly affects the older age group. And in the USA more people die of pneumococcal infection than of any other infectious disease that can be prevented by immunization. In Germany up to 22,000 people die of pneumonia each year, however it is difficult to know how many of these deaths are attributable to pneumococci and how many to influenza viruses, since in most cases no precise diagnosis is made. Moreover, pneumococci commonly cause secondary infection in elderly people whose defenses have already been weakened by influenza. Pneumococcal infections are also becoming increasingly common in HIV-infected people.

In developing countries the situation is different: there pneumococci are one of the most common harbingers of death in young children, accounting for up to 20% of all deaths in this age group. Worldwide, of the 1.6 million people who die of pneumococcal infection each year, one million are children under five years of age. And this despite the fact that an effective vaccine for use against these bacteria has been available for some time.

The vaccine took a long time to develop. This is because the bacteria possess a polysaccharide (sugar) capsule that protects them from drying out in the air and from ingestion by cells of the immune system. Moreover, the immune system is unable to build up any lasting immunity against these polysaccharide components, nor, therefore, can such immunity be achieved by vaccination. Only by conjugation of the polysaccharide components to proteins was this made possible. Worldwide there are more than 90 different pneumococcal strains, each of which differs from the others in terms of the polysaccharides present in its capsule. Presently available pneumococcal vaccines contain seven to thirteen polysaccharide components and provide protection against 50 to 80% of the pneumococci that cause pneumonia in young children.

Meningococci, another group of pathogens, likewise colonize the mouth and throat region. From there, however, they can spread to other parts of the body. The most important illness that they cause is a form of purulent meningitis that sometimes spreads to involve the brain itself and often has an extremely high mortality rate. However, since they generally remain in the mouth and pharynx without causing illness and can survive only in humans, they merit discussion in this section. Like pneumococci, meningococci possess a protective capsule, and as in the case of pneumococci, it is only recently that a conjugate vaccine for use against them has become available. Meningococci can be classified on the basis of their polysaccharide components into thirteen types, of which five are responsible for the vast majority of illnesses. The conjugate vaccine contains four of the five most common types of this pathogen. As yet, however, no vaccine is available against type B, the type most commonly found in Europe and the Americas, because the polysaccharide component of this type cross-reacts with components of human tissue and consequently could trigger an autoimmune reaction.

Recently, however, a vaccine directed against a protein component of type B meningococci was developed.

Though the occasional press reports of outbreaks of meningococcal meningitis in schools in Western Europe tend to cause great alarm, the real meningitis belt is situated in Central Africa, where there are frequent outbreaks of epidemic proportions such as that of 1996/1997, in which 200,000 people were affected. In Burkina Faso alone more than 22,000 cases have been reported since the beginning of 2007 and some 1500 people have died from the infection since that time. In the past there were also many outbreaks of meningococcal infection among pilgrims making their way to Mecca. For this reason Saudi Arabia now requires pilgrims to produce evidence of vaccination. Conjugate vaccines are very expensive. The news that a major pharmaceutical company has announced its intention of making a multicomponent vaccine available cheaply for use in babies in African countries is therefore most welcome. As well as protecting against diphtheria, tetanus, whooping cough, and hepatitis B, the vaccine concerned protects against meningococci, which are widely distributed in Central Africa.

5.2 Diarrheal diseases and food poisoning

Diarrheal diseases

Vacations in exotic destinations often bring memorable experiences. Less pleasant, however, is the experience of falling ill with diarrhea while traveling. Yet this is precisely what happens to half of all tourists and business travelers from industrialized countries who take trips to developing countries. In turn, almost half of these cases of traveler's diarrhea are due to infection with certain strains of the bacterium *Escherichia coli*, or *E. coli* for short. Every fifth case of diarrhea is due to viruses, in particular noroviruses. And 10% of infections are due to protozoa such as *Giardia*. The remainder are due to other bacterial pathogens, in particular shigellae and salmonellae. Each year more than 250 million episodes of diarrhea leading to more than 600,000 hospitalizations and 3000 deaths occur in adults in the USA. An etiologic agent is identified in less than 10% of cases.

In almost all cases infection occurs via food or drink, the most common sources being contaminated foods such as fish, ham, egg salad, seafood, unwashed fruit and vegetables, unboiled water, and ice made from unboiled water. Cold snacks sold on the street are never to be recommended in developing countries. Even in Germany, however, where standards of hygiene are high, more than 200,000 people develop diarrhea due to the above-mentioned pathogens each year. Every individual who is infected with such a diarrheal pathogen becomes an additional source of infection. People suffering from severe diarrhea excrete hundreds of millions of, and in some cases up to one billion, microorganisms per day.

In developed countries the vast majority of cases of diarrhea, though unpleasant, are of no great consequence to the sufferer. To many people in poorer countries, however, diarrhea means daily suffering and not uncommonly brings death. Millions of people still have no access to clean water or adequate sanitary facilities. Worldwide, two million people die of diarrhea and its complications each year – approximately the same number as die from tuberculosis.

Careful washing of the hands with soap can reduce the risk of infection by almost half. Few reliable studies are available on the influence of hand washing on deaths due to diarrhea, however according to some estimates use of such basic means alone could prevent half a million to one million of the two million deaths due to diarrheal diseases that occur each year.

E. *coli* is actually a harmless inhabitant of our intestine which, precisely because of its harmlessness, is commonly used by scientists in experiments. As a result, more is now known about E. *coli* than about any other organism on Earth. We know many strains of this bacterium down to their last item of genetic information, we know by name most of the proteins – channels, pores, transporters, docking sites for molecules, and many more – that are present in its membrane, and we know under what conditions these proteins work. Researchers can rotate 3-D animations of the proteins in their computers, zoom onto every atom, and observe how E. *coli* changes when, for example, a single atom is replaced by another. We know a huge amount about how the proteins are formed, which genes encode the information for the production of which proteins, and where these genes are situated within the organism's genome, and in many cases we at least have theories as to where the organism obtained its genes from in the first place. On the basis of all this information we distinguish many different types of E. *coli*. Not all of these, however, are harmless constituents of the intestinal flora. Rather, E. *coli* can cause many diseases. As described earlier, certain types of E. *coli* have imported genes from related pathogenic microorganisms. Many of these genes confer abilities on E. *coli* that can make it a dangerous aggressor. Such pieces of genetic information are also known as virulence genes. An example of this is provided by enterotoxin-forming strains of E. *coli* whose toxin can cause gastric and intestinal spasms as well as severe diarrhea sometimes accompanied by vomiting. The illness with the colorful name "Montezuma's revenge" is in most cases simply an infection with such ETEC (*enterotoxigenic E. coli*). As with most pathogens that cause diarrhea, humans infect each other via contaminated food. And a tiny dose of organisms is sufficient to cause a severe bout of diarrhea.

A number of other intestinal bacteria are prevalent in many countries. And most people are still familiar with the names of the associated illnesses, especially those – such as *dysentery* – that have been passed down from earlier times. The term "dysentery" is generally understood to mean diarrhea caused by a group of bacteria known as shigellae. Each year about 14,000 cases of shigella infection are reported in the USA, however the real number may be closer to 250,000. In Germany about 1000 people contract dysentery each year, many bringing it home from trips abroad. Worldwide, dysentery due to shigella is a major health problem, affecting about 160 million people annually. Of these people, about one million – mostly children in developing countries – die as a result of the diarrhea, dehydration, and exhaustion.

Cholera is another name that most people know to refer to a disease characterized by severe diarrhea. The responsible organism, *Vibrio cholerae*, completely disrupts the function of the intestinal villi (the myriad thread-like projections of the mucous membrane that lines the small intestine), interfering with the metabolism of the intestinal cells in such a way as to cause these cells to pump water and electrolytes into the intestinal lumen instead of drawing water, electrolytes, and nutrients out of the intestinal lumen and into the intestinal wall and the blood, as is – at least in most circumstances – their normal function. In this way cholera sufferers often lose more than five liters of fluid per day. Because of this enormous loss of water, the proportion of patients who die is higher than in other forms of diarrhea. In the mid-19th century Europe was ravaged by a number of cholera epidemics. Since then, however, the disease has become rare there. In developing countries the situation is quite different. In 1961 a new epidemic began, spreading in waves from Indonesia to East Asia, the Soviet Union, Iran, and Iraq. In 1970 it reached

West Africa, a region whose inhabitants had not had to deal with cholera for more than 100 years. The disease then spread throughout the entire African continent and outbreaks still occur there regularly. In 1990 cholera appeared in Latin America, once again in a region that had been free of the disease for 100 years. Where the necessary medical care is available, only about 1% of cholera patients die. Unless adequate amounts of fluid and electrolytes are given, however, about half of those affected soon succumb to the disease. As with other diarrheal diseases, clean water and basic hygienic measures are the keys to combating cholera.

The genetic information that codes for the production of the disease-inducing properties of *Vibrio cholerae* is held within the bacterium in a separate ring of genetic material known as a plasmid. And the bacterium can transmit this genetic information as a complete pack, so to speak, to related bacteria by exchanging entire plasmids with such bacteria. The recipient bacterium then treats these fragments of genetic material as if they were its own, reading the genetic information and passing it on to its protein factory, where it is translated to produce the virulence factor.

Protozoa – or more specifically *amoebas* – are also commonly responsible for diarrheal disease. In Germany up to 5000 cases of diarrhea are caused by these protozoa each year.

However, the top positions among the pathogens that cause diarrhea in many EU member states are occupied by viruses, principally noroviruses and rotaviruses. Noroviruses in particular have often made headlines in recent years. Occasional deaths, especially in old-age homes, have caused alarm. Many parents have become worried about their children, though noroviruses mostly infect adults. The diarrhea that they cause is extremely unpleasant, occurring within a few hours and being accompanied by profuse vomiting. In most cases the nightmare is over within three days. In Germany alone, more than 60,000 people develop norovirus infections each year. More important for infants and children are rotaviruses: worldwide, half a million to one million children die from diarrhea caused by these viruses each year. In the USA about 55,000 children are hospitalized each year because of diarrhea due to rotaviruses. It seems likely that almost all children suffer at least one rotavirus infection before their third birthday. Here again, the strategy for prevention is almost banal in its simplicity: clean water and the most elementary hygienic measures would prevent most deaths from occurring. In addition, a vaccine against rotaviruses is now available.

Food-induced diseases

What diarrheal diseases due to inadequate hygiene are to developing countries, food-borne diseases and food poisoning are to industrialized countries. And the costs of food-borne diseases are enormous, exceeding 50 billion euros per year in industrialized countries, even without considering the scandals surrounding the redating of batches of rotten meat and the sale of thawed and refrozen products that have rocked Europe in recent years. Some of the foods that we buy are incredibly cheap. The availability of meat and milk at such prices is due to intensive livestock farming and industrialized food processing techniques. At the same time, however, it is clear that in animal houses in which hundreds, if not thousands, of cattle or pigs are packed together in an extremely restricted space and in slaughterhouses in which as many as 10,000 chickens are slaughtered per hour or 10,000 pigs slaughtered per day, certain pathogens can spread like wildfire. As a result, animal infections often move from one animal house to another with amazing speed – and not uncommonly have to be dealt with by means of mass slaughter of the herds concerned. A recent example is an outbreak of swine fever that occurred in

North Rhine-Westphalia in March 2005. Following this in the biggest of the German states, the fact that over a period of only a few weeks more than 100,000 pigs were then slaughtered purely as a precaution passed almost unnoticed. Swine fever happens to be completely harmless to humans, however a number of other infectious diseases of animals can indeed pose a danger to humans. The BSE crisis – though it now appears to have been exaggerated to a sometimes hysterical extent during the confusing early period when perspective was lacking – showed among other things, how the way we farm animals can even give rise to completely new infectious diseases. Industrial processing of meat can result in microorganisms being rapidly passed from a single infected animal to an entire batch of a product. This can occur, for example, with the bacterium *Listeria*, which can be a problem in the production of minced meat for hamburgers, and with salmonellae in egg products.

It is estimated that in the USA alone about 80 million people suffer food poisoning each year. Of these, some 350,000 require hospital treatment and up to 7000 die. All too often such illness is attributable to incorrect or at least careless handling of meat in the kitchen. In the fall of 2006, following the occurrence of a number of outbreaks of salmonellosis with some fatalities, the German Federal Institute for Risk Assessment even issued an official warning about careless handling of poultry meat in the kitchen. According to this warning, poultry meat should be thawed without its packaging in the fridge and always – even when being grilled – cooked right through. Wooden chopping boards are unsuitable for cutting poultry meat because they are difficult to clean. In addition, the meat should be washed before being prepared, the kitchen equipment used to prepare the raw meat should be placed in the dishwasher immediately, and hands must be washed with warm water and soap between the individual preparation stages. All too often we are lulled into a false sense of security, believing that goods bought at the supermarket reach our table in sterile condition. But the truth is that for as long as the yolk of a fried egg remains liquid, microorganisms can survive in it.

Contamination of foodstuffs also has a substantial economic impact. In 1989 an enormous batch of meat in the USA became contaminated with *Listeria*. The resulting recall of the affected hamburgers cost the producer 76 million dollars alone, while loss of sales led to an additional loss of 200 million dollars.

Which microorganisms are the principal culprits here? We distinguish between more than 2500 closely related types of salmonella. *Salmonella typhi* and *S. paratyphi* are the agents responsible for typhoid fever and paratyphoid fever, respectively. In these serious illnesses the pathogen attacks the body's internal organs. Most other types of salmonella cause diarrhea, stomach upsets, dizziness, and vomiting. They are transmitted via contaminated foods and unclean water. Use of contaminated product batches from the foodstuffs industry regularly leads to outbreaks of salmonella infection in communal catering facilities such as canteens and nursing homes. The principal reason for this is the widespread presence of salmonella in intensive farming, especially poultry farming. At present, according to the German Federal Institute for Risk Assessment, every sixth broiler chicken flock and roughly every third laying hen farm in Germany is infected with salmonella. Compared with other European countries, Germany does not score well in this regard. In a number of Scandinavian countries the figures are considerably lower. Pork, too, is often contaminated with zoonosis pathogens. Outbreaks of illness in humans are no rarity, especially in summer. On average, salmonellae make 50,000 people in Germany ill each year, while Australia records more than 100,000, the UK more than 300,000, and the USA two to four million cases of diarrhea due to salmonella each year. The situation is made worse by the fact that antibiotic resistance has increased greatly: in Germany, for example, the proportion of these organisms that are resistant to antibiotics jumped from about 1% in 1990 to 17% in 1996 (see also Section 6.2).

An even more common cause of diarrhea in the industrialized world is the bacterium *Campylobacter jejuni*. In industrialized countries 5 to 15% of all cases of diarrhea are due to campylobacters. In Germany at least 60,000 cases occur annually and in the USA this pathogen strikes approximately two million times each year. Here too, some infections are life-threatening. Rarely, campylobacter infections can trigger Guillain-Barré syndrome. In this condition the affected person develops a severe form of paralysis but generally recovers completely within a few weeks. In older patients, however, the muscular weakness may persist for months and sometimes proves fatal. Like salmonellae, campylobacters are widely distributed in the animal kingdom, hence infection occurs mostly via contaminated food, typically inadequately cooked or raw meat.

Finally, staphylococci are another common cause of food poisoning. Though these bacteria are known mostly as pyogenic (pus-forming) organisms, many also produce a range of toxins that can cause vomiting and diarrhea. The difficult thing about them is that the toxin is heat-resistant and consequently unaffected by heating of the food in which it is present. The typical scenario of infection is a food processing worker with a suppurating wound on his hand who makes sweets, for example. The sweets are left standing for some time and during this time staphylococci are able to reproduce. Killing the microbes by subsequent heating of the sweet is of no avail, since by that time the microbes have already produced plenty of toxin. Staphylococci are responsible for around 250,000 cases of food poisoning per year in the USA alone.

Animal husbandry is a source not only of microbes that cause diarrhea, especially salmonellae and campylobacters, but also of influenza viruses. These are the subject of a separate section of this book. The whole problem is even more complex in that mass use of antibiotics in animal husbandry is partly responsible for resistance to antibiotics that are used in humans. This problem is referred to in detail elsewhere in this book

5.3 Children's diseases: far more than just a difficult start

At the time of soccer's 2006 World Cup in Germany health experts warned US citizens against traveling to Germany wide-eyed and unprotected. The Pan American Health Organization, the World Health Organization's regional office for the Americas, issued an official announcement urgently advising soccer fans to have themselves immunized – against measles – before traveling to the great sporting event. The reason for this was that in 2006 this almost forgotten children's disease had flared up strongly in Germany: whereas in 2005 only 780 cases had been reported, in 2006 the number of cases jumped to 2300. The vast majority of affected children and adults either had not been immunized or else lacked sufficient immunity. This outbreak of measles shortly before the World Cup is symptomatic of the way in which children's diseases are now perceived by many people in Western countries: thanks to the availability of vaccines and the systematic use of these over many years, these scourges of yesteryear have been largely overcome and accordingly have faded from our consciousness. This has resulted in a complacency about immunization which is now manifesting itself above all in children as insufficiently frequent booster immunizations against measles and low rates of uptake of the hepatitis B vaccine.

So it is that even in the industrialized world there are occasional outbreaks of "old acquaintances" such as measles. For example, by the middle of 2008 more than seventy cases of measles had been reported in at least seven states of the USA, more than in any other recent year. In 2006 the UK witnessed the biggest measles outbreak since 1988, with about 300 cases. One upshot of these outbreaks is that many people have once again been made aware of the threat that these diseases pose – as when, for instance, it became known that in the period between 2003 and mid-2007 a total of seventeen children and adolescents in Germany had died from complications of measles.

In general, the term "children's diseases" is used to refer to highly infectious diseases that can spread like wildfire in an unprotected population. On the other hand, a person who has once suffered a children's disease acquires a very prolonged, and in many cases lifelong, immunity to that disease. In general, therefore, adults are far less commonly affected by these diseases than children. However, adults who have no immunity because they have neither been vaccinated nor suffered the disease as a child are more or less as susceptible to the disease as are children. Apart from measles, the most important children's diseases that are due to viruses are chickenpox, mumps, rubella, and poliomyelitis, while the most important ones due to bacteria are diphtheria, tetanus, and pertussis. Effective vaccines are now available against all these pathogens. This gives cause for hope that in the next few years we may be able to eradicate some of them from the face of the Earth.

In the case of poliomyelitis this objective has almost been achieved, though recently there have been some setbacks. The clinical picture of poliovirus infection is marked by the development of paralysis, especially of the arms and legs, at a relatively early stage. In some cases the respiratory muscles are also affected, leading potentially to death from respiratory paralysis. Not uncommonly, however, poliovirus infection occurs without causing the classical signs and symptoms. Such 'silent' infections result in a natural immunity that provides protection against reinfection. This natural immunity is absent in Europe because the poliovirus has been largely eliminated from that part of the world. Even in Europe, therefore, immunization is still necessary – and will remain so until the virus has been completely eradicated.

At one stage the World Health Organization was in fact predicting that poliomyelitis would be eradicated by 2005. The prospects seemed excellent: whereas in 1988 a total of 350,000 cases of polio had been reported worldwide, by 2001 this figure had been reduced – thanks to massive immunization campaigns – to 483 cases. Shortly thereafter, however, polio immunization was suspended in the border zone between Nigeria and Niger because absurd rumors had been circulated there that it caused sterility and transmitted AIDS. As a result, it was not long before polio outbreaks began to occur again in that region: 430 cases there compared to 83 in all the rest of the world. And the disease advanced rapidly. The virus spread from the Niger-Nigeria border region to a number of other African countries and from there to Asia. In August 2005, 225 cases of polio were reported in Indonesia and in May 2006, 34 cases were reported in Namibia. By the end of 2006 statisticians had counted 1874 cases worldwide. At present cases of poliomyelitis are being reported regularly in Pakistan, India, Nigeria, and Afghanistan and sporadic outbreaks are occurring in twelve other countries.

Despite vaccination and the fact that the poliovirus can reproduce only in humans, the disease is proving very difficult to eradicate. This is because the virus is excreted in the feces and can then survives for a long time in the environment, e.g. in sewage.

Measles immunization programs have shown similarly encouraging results which, however, have been repeatedly marred by setbacks. Measles viruses cause fever, cold-like symptoms, and coughing that not uncommonly progress to severe inflammation of the meninges (the membranes that envelop the brain) and the spinal cord and then often lead to death. In addition, measles impairs the function of the immune system and for this reason is often followed by complications such as pneumonia and diarrhea of bacterial origin. More than half of all deaths due to measles occur as a result of pneumonia that develops when bacteria colonize the weakened organism. Statistics show that as recently as the early 1960s, 135 million cases of measles and more than six million deaths from measles occurred each year. The introduction of measles immunization soon met with extraordinary success: by 1987 the annual number of deaths had been reduced to 1.9 million. Three years later this figure had been reduced by more than half, to 875,000 deaths, while in 2005 only 350,000 people, mostly children, died as a result of measles. In 2006 the number of deaths due to measles fell further to less than 250,000. About 60% of these deaths occurred in Africa and 25% in Southeast Asia. It therefore comes as no surprise to learn that almost all the countries with the lowest measles immunization rates are in Africa.

In principle, the chances that measles can be completely eradicated seem to be reasonably good. The pathogen has no animal reservoir and can be transmitted only between humans. Epidemiologic studies have shown that in order to survive the virus requires at least a quarter of a million susceptible people in close contact with each other. Moreover, the fact that measles viruses scarcely change over time made it possible to develop a highly effective vaccine. On the other hand, the high infectiousness of the virus necessitates a very high level of immunization coverage. This is a precondition for the goal of eradicating, or at least almost completely overcoming, this children's disease by the year 2015.

5.4 HIV/AIDS

A quarter-century in the life of a virus

When the stage fright becomes unbearable shortly before the curtain rises, the men of the world's oldest gay men's chorus say to each other, "I sing for two." And wherever the San Francisco Gay Men's Chorus (SFGMC) sings, there is a red rose on the stage. "When SFGMC performs, we are more than the men that appear on stage," says the chorus's website. For each man standing, one chorus member has died of AIDS. Since the chorus was formed at the end of 1978 as a powerful voice for the rights of homosexuals, the list of deceased members has risen to over 250, more than the number of active members. There were times when the chorus lost a member a week. As reported by the *San Francisco Chronicle*, at every rehearsal during the 1980s and early 1990s there were announcements about who was in which hospital room and when the next memorial was scheduled.

Nowadays, continued the report, when another name is added to the list of deceased members, new members in their early twenties may whisper to one another that they've never known anybody with AIDS or HIV. At least in highly industrialized countries, the disease has now lost most of its horror. AIDS is no longer a death sentence. As the SFGMC approaches its thirtieth anniversary, it is experiencing a previously unknown sense of normality.

The history – now filmed – of the SFGMC tells part of the story of the epidemic of HIV/AIDS that began to appear between 1979 and 1980. At that time the public health departments of San Francisco and New York reported the strikingly frequent occurrence among homosexuals and drug addicts of illnesses that they had scarcely ever seen before. Firstly there were infections with the extremely rare microbe *Pneumocystis jirovecii*, which causes illness only in immunocompromised individuals, and secondly there were cases of an extremely rare form of cancer, Kaposi's sarcoma. *P. jirovecii* is now known to be a fungus that becomes established in the lung cells of immunocompromised individuals, where it causes a type of pneumonia characterized by a dry cough and fever. Kaposi's sarcoma is a malignant tumor whose development is triggered by a change in vessel walls. The characteristic reddish to brownish spots appear on the skin and mucous membranes and often also in the lungs and intestine.

These cases were first described by the USA's Centers for Disease Control (CDC) on June 5, 1981. It was soon discovered that both these conditions were associated with a previously unknown form of immunodeficiency. This led to the disease being named "acquired immunodeficiency syndrome" (AIDS). Only two years later the responsible microorganism was identified – a retrovirus that likewise was soon given a name: "human immunodeficiency virus" (HIV). Some years later researchers identified a second, similar, pathogen that was found mostly in patients from West Africa. Since then the first of the two viruses to be identified has been known as HIV-1 and the second as HIV-2. The former proved to be far more virulent than the latter and soon spread throughout the entire globe.

In the years since then we have learned an extraordinary

amount about HIV and AIDS; never before has so much information about a single pathogen and the disease that it causes been acquired in so short a time. There is no doubt whatsoever that HIV is the cause of AIDS, though the clinical picture of the disease is somewhat difficult to describe. As the name indicates, the disease is a deficiency of the immune system. Most patients die from other diseases. In industrialized countries most deaths result from infections with opportunistic organisms such as the fungi *Candida*, *Cryptococcus*, and *Pneumocystis*, which cause illness only in people with reduced resistance or wounds or illnesses. In developing countries tuberculosis occupies first place as an additional, and in many cases fatal, infection in AIDS patients. We will deal with the pernicious liaison between these two diseases later. The box "Chronology of a new plague: HIV/AIDS" summarizes the history of the disease to date.

The present situation

At present about 30–40 million people are living with HIV. In 2006 alone between four and five million people became infected with the virus. Since the disease was first described over twenty-five years ago HIV has claimed 25 million victims. An epidemic of unbelievable proportions has occurred, and its end is nowhere in sight.

Everywhere in the world the number of people who are living with HIV is increasing. Particularly dramatic increases have been seen recently in Eastern Europe, East Asia, and Central Asia. An estimated four million people in India and one million people in China are now infected with HIV. In Russia likewise, probably one million people are now living with HIV.

Whereas the wave of HIV infection that is presently occurring in China and Russia started among drug addicts and, especially in China, was propagated by high-risk sexual practices, the infection is now becoming increasingly common in women, especially prostitutes, in both of these countries and also in India. In Moscow more than 10% of drug addicts are HIV-positive, while in the Indian city of Mumbai more than half of all prostitutes are infected.

As before, however, the most dramatic situation is to be found in sub-Saharan Africa: in this region two-thirds of the world's HIV-infected people live and almost three-quarters of all deaths are due to AIDS. The disease has reduced life expectancy in the region to such an extent that, for example, a child born today in Botswana is unlikely to live beyond the age of thirty years.

Fig. 8 shows the number of people with HIV/AIDS, Fig. 9 the number of newly infected people, and Fig. 10 the number of AIDS-related deaths in the various regions of the world in 2006.

Germany, by contrast, has the problem of AIDS under some degree of control. In 2007 the public health authorities reported more than 2700 new cases of HIV infection. Though this is the highest number to be reported since detailed records began, a large part of the increase is attributable to better statistics and more and simpler tests. In 2006 a total of 620 HIV-infected people developed AIDS and about 600 people died of AIDS. It is estimated that about 56,000 people in Germany are currently living with HIV. Though this figure may seem high, the availability of ever more effective drugs means that many of these people are now able to lead a reasonably normal life. In the USA new diagnoses of HIV among men having unprotected sex increased in 2001 from almost 18,000 to 19,600 individuals (53%) in 2005. In the UK, between 2001 and 2006, the number of new diagnoses doubled signaling an ongoing HIV epidemic. Nevertheless, although AIDS is now treatable, it is not yet curable, therefore the number of infected people will continue to rise. There is no reason to sound the all-clear: the side effects of the drugs are

Chronology of a new plague: HIV/AIDS

About 1945: A virus that will later become known as HIV
      is transmitted from chimpanzees to humans.

1959: A man in present-day Congo is found to have HIV
      in his blood.

1981: The first AIDS cases are reported in homosexual
      men and drug addicts in the USA.

1982: Clinical definition of AIDS established; routes of
      transmission elucidated.

1983: In France, Dr. Luc Montagnier (Nobel Prize 2008)
      isolates a virus from patients that will later be
      named HIV. First signs of an AIDS epidemic among
      heterosexuals in Central Africa.

1985: AIDS is a worldwide problem. First HIV antibody
      tests developed in USA and Europe.

1986: A second human immunodeficiency virus (HIV-2) is
      discovered in the blood of West African patients.

1987: AIDS becomes the first disease to be discussed in
      the UN General Assembly. The first drug for use
      against AIDS (AZT = azidothymidine) is approved
      in the USA.

1988: The World Health Organization proclaims December
      1 as World AIDS Day. In sub-Saharan Africa as many
      women as men are now infected with the virus.

1990: One million children are now orphans due to AIDS.

1994: First possibilities for preventing transmission from mother to child.

1995: Increasing HIV outbreaks among drug addicts in Eastern Europe.

1996: Brazil becomes the first relatively poor country to offer people antiretroviral combination therapy via the public health system.

1998: Thirty-nine pharmaceutical companies file a suit against the sale of cheap AIDS drugs in South Africa, but then withdraw the suit.

1999: AIDS discussed for the first time in the UN Security Council.

2000: The UN Millennium Development Project sets a goal of stopping the spread of AIDS by 2015.

2001: The UN General Assembly declares AIDS to be a global catastrophe.

2002: Global Fund to Fight AIDS, TB and Malaria established.

2003: US President George W. Bush promises 15 billion dollars for the fight against AIDS.

2005: 1.3 million people in developing and rapidly industrializing countries receive treatment.

2006: Forty million people are living with HIV; 25 million people have died of AIDS.

2007: Twenty-one drugs from four drug categories approved for use in HIV/AIDS; increasing problems with resistant viruses in industrialized countries.

Adults and children with HIV in 2007

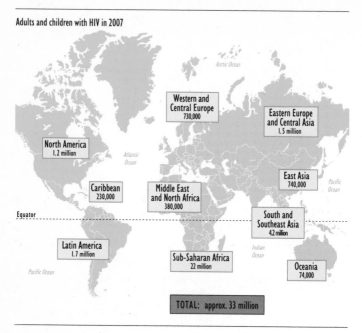

Figure 8   Number of adults and children living with HIV in the various
            regions of the world in 2007
            Source: UNAIDS (2008)

substantial and patients have to put up with various symptoms.
What's more, transmission of multidrug-resistant viruses is
increasing, making treatment more difficult.

In 2007 the World Health Organization and UNAIDS cor-
rected the numbers for AIDS based on new methods of estima-
tion. Accordingly, the number of individuals newly infected with
HIV is estimated in the order of 2.7 million (with a range of
1.8–4.1 million) and the number of individuals living with HIV
in 2007 was estimated at 33 million (with a range of 30.6–36.1

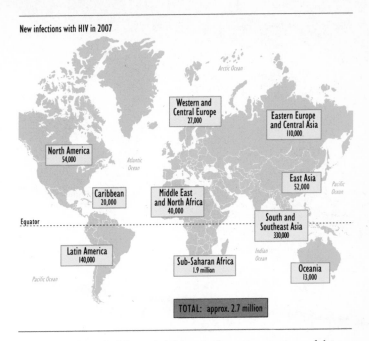

New infections with HIV in 2007

Western and Central Europe
27,000

Eastern Europe and Central Asia
110,000

North America
54,000

East Asia
52,000

Caribbean
20,000

Middle East and North Africa
40,000

South and Southeast Asia
330,000

Equator

Latin America
140,000

Sub-Saharan Africa
1.9 million

Oceania
13,000

TOTAL: approx. 2.7 million

Figure 9   Number of adults and children in the various regions of the
world newly infected with HIV during 2007
Source: UNAIDS (2008)

million). These figures are lower than previous ones, indicating
the difficulty of making precise estimations. Rates may be over-
estimated by organizations interested in raising awareness about
HIV/AIDS and under-estimated by states interested in minimiz-
ing publicity about this problem.

From a virus's bag of tricks

The insidiousness of HIV is due principally to three characteristics of the virus:

Adult and child deaths from AIDS during 2007

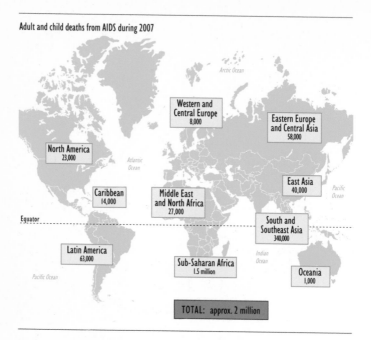

Figure 10 Adult and child deaths from AIDS in the various regions of the
world in 2007
Source: UNAIDS (2008)

1. The virus attacks CD4 T helper cells, which are the central
   control centers of the immune system and are required
   for every immune response – whether this be in the form
   of CD8 T killer cells, which destroy virus-infected cells;
   macrophages, which eliminate bacteria, fungi, and parasites;
   or B lymphocytes, which produce antibodies. Once CD4 T
   cells have been eliminated, sooner or later the entire immune
   system will inevitably break down.

2. The virus is able to transform itself endlessly, forever altering

its appearance in the patient. As is so often the case in nature, the secret of this ability is a reproductive machinery that is prone to error: as a retrovirus, HIV stores its genetic information in the form of RNA that has to be transcribed into DNA in cells. This occurs in a very imprecise fashion. Since, in addition, HIV reproduces itself very rapidly, it mutates continuously and in this way keeps on sidestepping the immune response. Because of the high mutation rate, the likelihood that resistance will develop is also far higher with HIV than with many other pathogens.

3. The genetic information of the virus is incorporated into the host's cells in the form of a "provirus". This allows the pathogen to survive in cells with a long lifespan and means that even in the presence of effective antiviral therapy it never disappears completely from the body.

A virus by itself does not make an illness

In line with the central importance of CD4 T helper cells, the number of these cells is crucial in determining progression of the disease. In healthy people more than 1000 of these cells are present in each microliter of blood, while the number falls continuously with increasing severity of illness. Roughly speaking, four stages of the disease are distinguished. In the acute phase, i.e. in the first few days after the virus has found its way into the body, many people are scarcely aware of the infection, whereas in others it shows itself as generalized weakness, fever, and swollen lymph nodes. Many of those affected develop diarrhea or difficulty in swallowing. Only rarely do people visit a doctor on account of these symptoms, and if they do, they are probably not thinking of HIV. Then things settle down for a while. This asymptomatic stage passes more or less without evidence of illness, though the presence of viral

RNA can be demonstrated, i.e. the person is now HIV-positive. Next comes the stage of chronic infection, in which the
infected persons occasionally feel unwell, their skin or mucous
membranes change, they develop gastrointestinal disturbances,
and many also develop neurologic problems. Finally, AIDS is
said to be present when opportunistic organisms strike. At this
stage microorganisms that are unproblematic for healthy people
rapidly cause life-threatening illness. When the concentration of
CD4 T helper cells falls below 200 per microliter of blood and
superinfection with other microorganisms occurs, the person
suffers from severe or full-blown AIDS.

The route of the virus – and how it can be interrupted

As every child now learns in school, AIDS is transmitted via
bodily fluids. The most important route is via sexual intercourse.
The risk of transmission is particularly high in sex that results in
bleeding, e.g. anal or vaginal penetration, notably in the presence
of sexually transmitted diseases that cause mucosal injury. On
the other hand, contact that does not lead to exchange of bodily
fluid, in particular blood, does not pose a risk.

When correctly used, condoms prevent transmission of HIV
during sex with 80 to 90% certainty and also protect against
other sexually transmitted diseases. It is estimated that in Thailand the use of condoms prevented about 200,000 HIV infections
in the period between 1993 and 2000. Moreover, condoms – like
sexual abstinence – are cheap.

However, a precondition for successful use of condoms in
AIDS prevention is a high level of acceptance among the population. In this regard a great deal of education and convincing
still needs to be done, especially in sub-Saharan Africa. Not least
among the obstacles to regular use of condoms are men. The
prospects would therefore be far better if women were able to

take protective measures at their own initiative and completely independently of men. Much hope has been placed in foam and gel products that kill viruses in the vagina. Unfortunately, the first large-scale clinical trial of such a microbicidal gel for use against HIV turned into a fiasco when it was found that more women in the treated group had become infected with HIV than in the control group. Notwithstanding this result, more work should be done on this approach. Also interesting are studies according to which circumcised men are only about half as likely as uncircumcised men to become infected with HIV during sex with women. This is believed to be due to the accumulation of HIV in immune cells in the foreskin.

The second means of transmission of HIV is blood contact – via contaminated injection equipment in drug addicts, due to inadequate hygiene in hospitals, and via transfusions. Provision of clean needles and syringes to intravenous drug users can therefore make a significant contribution to limiting the spread of the disease. The AIDS epidemic in Eastern Europe and Asia was set in motion by widespread reuse and sharing of needles among intravenous drug users. In China about half, and in Eastern Europe considerably more than half, of HIV-infected patients probably acquired the disease in this way.

Reuse of injection equipment also still occurs in many hospitals around the world. Young children are often the victims. Appalling hygienic conditions of this kind are thought to have been responsible for an outbreak of HIV infection in a children's hospital in the Libyan port of Benghasi in which more than 400 children – fifty-six of whom have since died of AIDS – were infected. In May 2004 a Libyan court, on the basis of some bizarre reasoning, sentenced five Bulgarian medics and a Palestinian doctor who had worked at the hospital to death for deliberately infecting the children with HIV. Later the accused

were returned to Bulgaria, where their life sentences were immediately commuted and they were freed.

The safety of transfusion of blood products must therefore
be regarded as problematic in such countries. In Western Europe
and the USA the prescribed tests and precautions largely exclude
the possibility of HIV infection by blood transfusion and similar
procedures.

Attack on a killer

AIDS is now treatable, but not curable. Combination therapies
consisting of a number of antiretroviral drugs can effectively halt
viral replication but are unable to eliminate the virus. Where
possible, preference is now given to aggressive chemotherapy
(HAART: highly active antiretroviral therapy) consisting of a
number of drugs that enhance each other's action. This inten-
sification of antiretroviral therapy (ART) by combining up to
five drugs largely frees AIDS patients from the symptoms of the
disease (though so far often at the cost of substantial side effects)
and hinders the development of resistant viruses.

In poorer countries the goal of universal treatment of AIDS
patients is still made difficult not only by cultural barriers, but
above all by the high cost of the drugs. At present fewer than two
million people in poor countries, including one million in Africa,
are receiving ART. Fig. 11 shows the proportion of people with
HIV/AIDS in sub-Saharan Africa who received ART in 2007. If
nothing changes, millions of AIDS patients in developing countries will die from the disease over the next years purely because
of lack of drugs.

On the other hand, the side effects of AIDS therapy are
considerable, and the more people are treated, the greater the
number of resistant viruses that appear. This risk is particularly
pronounced in children who have carried the virus since birth

because their mother was HIV-positive. Such strains can be very difficult or even impossible to treat.

For example, the year 2004 saw the appearance in New York of a virus that was found to be resistant to nineteen out of twenty drugs. A man described in the newspapers as being in his mid-forties became severely ill within a few weeks of acquiring the extremely resistant virus at a "bareback" (unprotected anal sex) party.

Blockade at a number of levels

Just as antibiotics can attack bacteria at a number of levels, so too can drugs directed against HIV/AIDS attack the virus at a number of levels. For example, replication of the viral genome can be blocked by artificial genetic components. Thus, the action of reverse transcriptase, the enzyme that transcribes viral RNA into DNA, can be blocked by nucleoside analogs. The best-known drug of this class is azidothymidine (AZT), also known as zidovudine. Nevirapine likewise acts by inhibiting reverse transcriptase, though it is not a nucleoside analog.

Protease inhibitors attack the virus at the level of proteins by inhibiting the cleavage of a large protein molecule of HIV into its constituent subunits. This cleavage is a precondition for the formation of infectious viral particles. Saquinavir belongs to this group.

Uptake of virus into the host cell, i.e. fusion of viral particles with the host cell, can also be blocked by appropriate drugs. Enfuvirtide is one such fusion inhibitor.

Also promising are clinical studies with drugs that prevent cells from integrating the viral genes into their genome via the enzyme integrase. Raltegravir is one such integrase inhibitor.

Antiretroviral (ARV) treatment coverage

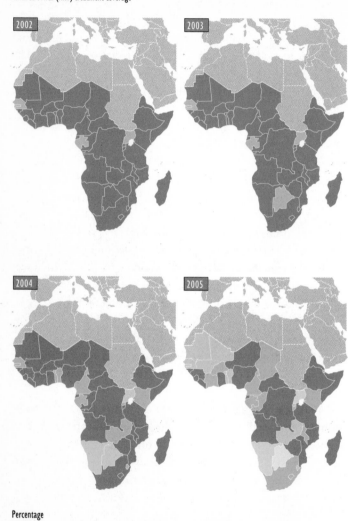

Percentage
75 – 100   50 – 74.9   25 – 49.9   10 – 29.9   less than 10

A question of cost

The question of how medicines can be made available to people in poor countries at affordable prices, and about the role of pharmaceutical companies in achieving this, has generated a great deal of public interest, especially in relation to AIDS. One result of this interest is that most people are now familiar with the term "generic", meaning a cheaper version of a brand-name drug. This discussion still has a long way to go.

Up to now some 300,000 AIDS patients in developing countries, equivalent to half of all treated AIDS patients in those countries, have been supplied with cheap generic medicines. Since 2001 Indian generic drug manufacturers have been producing AIDS drugs and offering them for sale at far lower prices than the major pharmaceutical companies. The patented drugs of the pharmaceutical giants cost between 15,000 and 20,000 US dollars per patient per year, whereas their generic equivalents cost less than 500 US dollars per patient per year. Almost equally importantly, the smaller companies produce the most effective combinations regardless of who originally developed and patented the drugs. No wonder, then, that the pharmaceutical giants are displeased. Via the World Trade Organization massive pressure was put on India to tighten its laws so as not to infringe the patent rights of the industrialized countries. At the same time, a group of thirty-nine pharmaceutical companies filed a lawsuit against the South African government aimed at safeguarding their patent rights for AIDS drugs and preventing the sale of these at reduced prices, though they later dropped the lawsuit. We will come back to this question later.

Figure 11   Number of AIDS patients in Africa receiving treatment 2007

Persons with HIV/AIDS receiving antiretroviral therapy in various regions of the world, in 2007. Number of people treated as a percentage of those who needed treatment. Source: WHO/UNAIDS/UNICEF (2008)

Severely ill from birth

During childbirth blood passes between the mother and the infant, therefore the risk that an HIV-infected mother will pass the virus on to the infant is extremely high. Timely drug treatment of the HIV-positive mother, delivery by cesarean section, and feeding of the neonate on milk products rather than breast milk can prevent HIV transmission in most cases. In industrialized countries these measures have succeeded in reducing the risk of mother-to-child transmission from 30 to 40% to 2 to 3%. In developing countries, by contrast, the situation remains tragic: less than one in every ten HIV-infected pregnant women is given the opportunity of preventing transmission of the disease to her infant. A third of infants who are HIV-positive at birth die in their first year of life and almost half die in their second. As well as having to cope with severe illness, those who survive often become orphans or half-orphans before very long.

Immunization

Though attempts to produce a vaccine against HIV/AIDS have been going on for years, the results achieved to date have been disappointing. Three basic approaches have been adopted:

1. Vaccines that stimulate the production of antibodies, i.e. humoral immunity: Ideally, immunized people would possess so many neutralizing antibodies that these would simply prevent HIV infection from occurring in the first place. For this to happen, high concentrations of antibodies would be required especially in the mucous membranes of the penis, rectum, and/or vagina. However, the clinical trials performed to date with vaccines that induce the production of neutralizing antibodies against HIV have failed, suggesting that this problem will not prove so easy to solve.

2. Vaccines that stimulate cellular immunity, in particular
   T killer cells: In this approach the aim is to implant HIV
   antigens in suitable vector systems that then induce an
   effective T killer cell response. The T killer cells would
   probably be unable to eliminate the virus completely, but
   might be able to keep the HIV concentration low. It is not
   certain whether such protection would be sufficient in the
   long term, since HIV suppresses the central control centers
   of the immune system, namely CD4 T helper cells, which are
   responsible for all other aspects of the immune response.

   A second problem with this type of vaccine is that given
   the virus's great ability to change, it is to be feared that
   sooner or later the virus will escape the immune response
   and start to reproduce again unchecked. Though not
   conclusive, the results of the clinical trials performed to date
   with this type of vaccine give little cause for optimism. One
   large-scale clinical trial with such a vaccine was terminated
   prematurely because the vaccine failed to show sufficient
   safety or efficacy.

3. Vaccines that stimulate broad cellular and humoral
   immunity: In view of the poor results obtained with the
   two types of vaccine referred to above, attempts are now
   being made to develop vector systems that stimulate both
   neutralizing antibodies and T killer cells and that cover
   the broadest possible range of viral antigens. In the search
   for neutralizing antibodies use is made of the most recent
   findings in structural biology and peptide chemistry.
   The aim is to identify target structures in HIV that are
   recognized by antibodies during infection. The attempt
   to stimulate T killer cells is based upon viral vectors that
   are known to induce a highly effective and very broad T
   cell response. Finally, attempts are being made to improve

effectiveness by means of special immunization schedules. For example, the vaccines used for primary and secondary (booster) immunization may be based upon different vectors that nevertheless express the same antigens. However, such approaches are expensive and time-consuming and are made even more difficult by the range of tricks that HIV has at its disposal. Even professional optimists therefore do not believe that an effective vaccine against AIDS will become available within the next ten years.

5.5 Tuberculosis: the white plague

Overview

Tuberculosis is the most tenacious of all infectious diseases. How many times has it been declared to have been defeated, only to bounce back, mostly with increased vigor? More people are dying from this disease now than ever before. And HIV/AIDS has made things even easier for the responsible bacterium, *Mycobacterium tuberculosis*. The figures are truly frightening: every third person in the world, i.e. about two billion people, are infected with this organism, also known as the tubercle bacillus.

Of those who are infected, 15 million are ill, and each year nine million new cases are added to the list. Almost two million people die each year from tuberculosis. Put more starkly, this means that every day 25,000 people become ill with tuberculosis and 6000 people die from the disease – a death every fifteen seconds.

The country with the highest absolute figures for tuberculosis is India, with about 1.1 million notified cases and about five million estimated cases. Next comes China, with just on half a million notified cases and approximately three times as many estimated cases. The worst affected region, however, is Africa, where of a total population of about one billion, 2.5 million people are ill with tuberculosis. The tuberculosis figures for the various regions of the world are shown in Fig. 12.

And the situation is made worse by the fact that the bacteria are becoming increasingly resistant to drugs. We now know not only mycobacteria against which the best drugs are ineffective (MDR = multidrug-resistant), but also strains against which scarcely any of the presently available drugs are effective (XDR = extensively drug-resistant). Around 50 million people are now infected with MDR strains of mycobacteria and each year half a million more people become infected. In 1993, when the World Health Organization recognized the problem and declared tuberculosis to be a global emergency, it was already almost too late. Nor is this disease going to leave us in peace in the new millennium. And unless we achieve decisive breakthroughs very soon, 100 million more people will become ill with the disease, and 20 million will die from it, in the first decade of this century alone.

The history of tuberculosis: fear over thousands of years

The first evidence of tuberculosis comes from 5000 years before Christ. Evidence of tuberculosis is also apparent in Egyptian mummies (ca. 2400 BC), and in a few cases the tubercle bacillus has even been demonstrated in mummies using molecular genetic techniques. Descriptions in the Bible and in Indian and Chinese texts thousands of years old underline the fear that this disease has inspired over millennia. The disease was first described in detail in the first and third books of the Hippocratic Corpus, the written work of a school of medicine centered around the Ancient Greek philosopher and physician Hippocrates (460–370 BC). It was here that the term "phthisis" (wasting away) was first used to refer to tuberculosis.

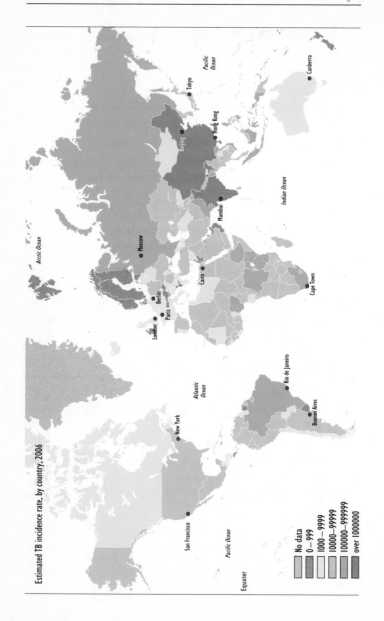

Estimated TB incidence rate, by country, 2006

No data
0 – 999
1000 – 9999
10000 – 99999
100000 – 999999
over 1000000

In a text entitled *Problemata* the Ancient Greek philosopher Aristotle (384–322 BC) considered the question of how the disease is transmitted: "Why do people who have come into contact with someone who is afflicted by phthisis, ophthalmia, or psoriasis become ill with these same diseases?" Girolamo Fracastoro (1478–1553), the great Renaissance scholar of contagious diseases, likewise foresaw the manner in which tuberculosis is transmitted. This is remarkable insofar as tuberculosis often appears only a long time, in some cases years, after contagion. Certainly, Fracastoro recognized both the transmissibility of the disease and the susceptibility of human beings as being critical factors – and did so at a time when the kings of France and England, utterly convinced of their own powers and to the great admiration of their subjects, celebrated the laying on of hands as a means of curing tuberculosis by the "royal touch." Between 1662 and 1682 King Charles II of England (1630–1685) laid his hand on more than 90,000 of his ill subjects – and if contemporary reports are to be believed, to great effect. Similarly, the ten-year-old Louis XIII of France blessed 800 tuberculous subjects on the occasion of his coronation. In this case, however, the blessing appears to have been less effective, since Louis XIII himself later developed the disease.

Scrofula and consumption: two clinical pictures of the same disease

Tuberculosis is more than just tuberculosis. On the one hand there is the otherworldly-looking consumption, which was accorded almost cult-like status in the Romantic period. On the

Figure 12 Estimated tuberculosis incidence rates in the various regions of the world in 2006

    Number of new cases of tuberculosis (all forms) per 100,000 population
    Source: World Health Organization (2008)

other hand there was the repulsive-looking victim of scrofula, with inflamed, ulcerated lymph nodes in the neck region. The latter form of tuberculosis is now rare, not least because it probably occurred as a result of a route of transmission that is now rare, namely ingestion of contaminated food or drinks, in many cases the milk of tuberculous cows. With the introduction of pasteurization and the successful rearing of tuberculosis-free herds of cattle the problem – and this form of tuberculosis – disappeared in most industrialized countries. In some parts of the world it still occurs (see box: "Tuberculosis rages among the lions of South Africa").

By contrast, pulmonary tuberculosis, or what used to be known as consumption, has now become the most common form of tuberculosis. It takes the form of a wasting disease in which the patient generally suffers from chronic fever, night sweats, and coughing. Later bloodstained sputum is produced. Although all organs of the body can be affected, the lungs are the organ most commonly affected. The agent is transmitted by droplet infection, i.e. by coughing and expectoration. On average, a patient with active pulmonary tuberculosis infects ten to fifteen other people per year in this way.

Infection and illness

The infectious etiology of tuberculosis was discovered by the Berlin physician and scientist Robert Koch (1843–1910). At Koch's time a third of all deaths in Berlin, Paris, London, and other European cities were due to tuberculosis. Robert Koch also originated the tuberculin test, in which the appearance of a moderately large reddish lump after the injection into the skin of products of the tubercle bacillus is taken to indicate either infection or successful immunization.

On the basis of mass screening using tests of this type, we

know that two billion people worldwide are infected with *Mycobacterium tuberculosis*. In some Eastern European countries, in Southeast Asia, and certainly in Africa, more than half the population is infected, while in certain "hot spots" almost the entire population is living with the tubercle bacillus. And there is nothing new about this. When a large number of cadavers were examined in Zurich at the beginning of the 20th century it was found that all the adult deceased had tubercle bacilli in their lungs – even those who had never been diagnosed with tuberculosis and those who had died of other causes. Nowadays a very similar situation prevails in the townships of the cities of sub-Saharan Africa, in the prisons and labor camps of Russia, and in the slums of the cities of India (see also box: "Funeral homes with bars: tuberculosis in Russian prisons").

Immunization

A vaccine for use against tuberculosis was developed by Albert Calmette (1863–1933) and Camille Guérin (1872–1961). In their honor it still bears the name BCG, for *bacille Calmette-Guérin*. It was developed as a means of protecting neonates born into a tuberculous family. The vaccine is still indicated in this situation today, as it prevents the most severe forms of tuberculosis in infants. And now that about four billion doses of it have been administered, its high degree of safety is well established. However, we now know that it does not provide protection against the most common form of tuberculosis, namely pulmonary tuberculosis in adults. For this reason scientists in many research centers are now working to develop new tuberculosis vaccines. Some of the candidates are now being tested in clinical trials. Three different strategies are being followed:

1. Substitutes for BCG: Attempts are being made to improve

Tuberculosis rages among the lions of South Africa

Among the most impressive experiences to be had in Africa are safaris in the Kruger National Park. Especially in the south of the park, visitors have a good chance of seeing lions basking in the sun. Even more exciting, of course, is to see how these magnificent predators work together to hunt their prey. In December 1998 we were confronted for the first time by a lion that was acting oddly. It was emaciated, mere skin and bones. It crept along sluggishly in front of us, unconcerned and apparently unaware. In response to my questions I was told that this was one of the countless lions in the southern part of the park that suffered from tuberculosis and was doomed to die. Since then 80% of the lions in the southern part of the Kruger National Park have become infected with *Mycobacterium bovis*, the organism responsible for bovine tuberculosis.

Bovine tuberculosis probably came to South Africa with the herds of cattle brought by the white immigrants particularly in the 18th century. In the early 1960s there was repeated contact between cattle herds and the native wild buffaloes of the Kruger National Park. The buffaloes caught the disease by droplet infection, just as humans do. By the time the National Park was fenced in, the disease was established in the buffalo herds. The disease spreads slowly among buffaloes and most infected buffaloes do not become ill at all – as is the case with humans. But precisely for this reason the buffalo herds in the southern part of the Kruger National Park became an enormous reservoir for the agent of bovine tuberculosis. The disease kills very few of the animals, however it weakens them, especially in the dry season.

This then takes the disease one step higher in the food chain. This is because lions avoid unnecessary exertion when hunting and therefore often choose a sick, and consequently weakened, buffalo as their prey. The leaders of the pride then get to eat the most desirable parts of the carcass. These include the lungs, which contain vast numbers of tubercle bacilli. As a result, these dominant lions soon become ill, typically with gastrointestinal tuberculosis, and rapidly become emaciated. The wasting away of the dominant animals then affects the composition of the pride, since younger lions now make premature attempts to take over the leadership. The rangers of the Kruger National Park have noticed a dramatic fall in the average age of animals in the pride, a shift in the sex ratio, and significantly fewer offspring. As a result, tuberculosis could upset the entire ecosystem of the park.

Other big cats such as cheetahs and leopards have also become infected. In August 1998 a cheetah weakened by tuberculosis dropped straight out of a tree onto a ranger and killed him, no doubt because it had become too weak to hunt wild animals.

At present around twenty-five of the approximately 2200 lions in the Kruger National Park die of tuberculosis each year. The park administration is largely powerless to deal with the problem. The only obvious solution would be to completely separate the northern from the southern part of the park with a fence so as to confine tuberculosis to the southern part. But elephants keep tearing fences down. And now experts are worried that bovine tuberculosis may also pose a risk to humans. In this scenario, buffaloes would transmit the disease to domestic cattle, which could then infect humans. Given the millions of HIV-infected, and therefore immunocompromised, people in southern Africa, this is a frightening prospect.

BCG by molecular genetic means. In one approach, additional antigens that are preferentially recognized by protective T cells are being inserted into the vaccine. In another approach, attempts are being made to alter the BCG vaccine in such a way that it induces a more powerful immune response. The idea is to stimulate not only the important T helper cells, but also T killer cells. Such vaccines would be administered shortly after birth in place of BCG. In order to win marketing approval they would need to be safer than BCG and/or provide more effective protection. Two vaccine candidates of this type have entered clinical trials.

2. Vaccines that improve the protection provided by BCG: The aim of this second strategy is to strengthen the immunity provided by BCG. To this end subunit vaccines are composed which on the one hand contain antigens that are preferentially recognized by T cells and on the other hand strengthen the T cell response by use of new adjuvants. Such T-cell stimulating adjuvants were developed only a few years ago. So far no vaccine candidate of this type provides better protection than does BCG, however when administered after BCG such vaccines can improve the protection provided by BCG. This strategy is known as heterologous prime/boost immunization. A number of subunit vaccine candidates of this type are now at the stage of clinical trials.

3. Combination of new vaccine candidates for heterologous prime/boost immunization: The above two immunization strategies could of course be combined. The best results could probably be achieved by combining prime immunization using a new, improved BCG with booster immunization using a subunit vaccine, however the only means of finding this out would be via extensive clinical trials. The individual vaccine candidates would need to

Funeral homes with bars: tuberculosis in Russian prisons

In Russia tuberculosis has its own special breeding ground: prisons. There people live in an extremely confined space under appalling hygienic conditions. This is an ideal environment for the spread of the tubercle bacillus.

Over the last decade of the 20th century the number of people sent to prison in Russia increased enormously. For every 100,000 inhabitants there were up to 1000 prisoners, i.e. about 1% of the population. Soon the labor camps and prisons were packed with about one million inmates. According to a report by Radio Free Europe, inmates of Russian prisons are confined within an area of about 2.5 square meters – about the size of a coffin. Approximately one prisoner in ten living in this tiny space, i.e. about 100,000 people in all of Russia, suffer from active pulmonary tuberculosis, a highly infectious form of the disease. From the prisons the victims spread the disease to the population at large, since every year the prisons receive about 300,000 new inmates and – because of the overcrowding – roughly the same number are released. On this basis it can be calculated that every year about 30,000 people with active pulmonary tuberculosis step through the gates of Russian prisons into the world outside. At present the rate of tuberculosis in Russian penal institutions is about forty times that outside of the bars.

The problem is compounded by the emergence of multidrug-resistant organisms. At present a quarter of affected ex-prisoners have MDR-TB (multidrug-resistant tuberculosis). And sooner or later the XDR-TB (extensively drug-resistant tuberculosis) that is presently spreading in Africa and other regions is sure to appear. On top of this, Russia is now experiencing an AIDS epidemic. At present about 35,000 Russian prisoners are living with HIV and are therefore at particular risk of contracting tuberculosis.

be tested first, and only then the combinations. A new immunization regimen for tuberculosis is therefore not to be expected within the next ten years at least.

Chemotherapy and the development of resistance

Tuberculosis is treatable, however the treatment is no small matter. The pathogen grows very slowly. In active pulmonary tuberculosis the number of organisms becomes enormous: the lungs of an adult human with active pulmonary tuberculosis can contain far more than a trillion (10^{12}) bacteria. And conventional antibiotics don't work.

The treatment of tuberculosis calls for special drugs. At least three, and in most cases four, antituberculous drugs have to be administered simultaneously over a period of six months. Not uncommonly treatment fails because of lack of compliance on the part of the patient or incorrect use of the drugs. In order to improve the treatment of tuberculosis, the World Health Organization recommends the DOTS ("directly observed treatment, short course") strategy. This requires patients to take their medicines in the presence of medical or paramedical personnel. Since 1995 more than 15 million people have received DOTS. In areas in which effective diagnosis of tuberculosis occurs and resistant organisms are rare, the approach has proved successful.

Nevertheless, because of the long duration of treatment and the severe side effects of the drugs, many patients stop treatment prematurely. This is referred to as inadequate compliance. The result is a dramatic increase in MDR-TB. In patients with MDR-TB even DOTS can easily fail. What's more, in many such patients DOTS then favors the development of resistance. This is because the classical drug cocktail allows monoresistant organisms to become multiresistant very quickly. Once this happens, only second-line drugs will help.

The incidence of MDR-TB has increased alarmingly, especially in patients who have had an unsuccessful course of treatment. Of the twenty countries with the highest rates of MDR-TB, fourteen are situated in Europe or Central Asia. In Kazakhstan more than half of all tuberculosis sufferers have MDR-TB. Next comes Estonia, with over 45%, then the Tomsk region of Russia, with over 43%. In industrialized countries resistant strains create mostly financial problems, as they are 100 to 1000 times more expensive to treat than susceptible strains. In developing countries this increase in cost means that most patients no longer have access to effective treatment.

Even more alarming is the increase in the number of cases of XDR-TB, effective treatment of which is unavailable even to the wealthiest of people. In early 2008, XDR-TB had been demonstrated in forty-six different countries. XDR-TB is now present not only in developing countries, but also in the USA and many EU countries. In the USA 4% of MDR-TB cases are now extensively resistant, while in Latvia one in five MDR-TB cases is extensively resistant.

The World Health Organization has dubbed XDR-TB "Ebola with wings" and considers the pathogen to be as dangerous as the H5N1 virus of avian influenza. In 2006 the world watched on in horror when XDR-TB broke out in a group of HIV-infected patients in a hospital in the South African province of KwaZulu-Natal. Of fifty-three patients, fifty-two died. It must be feared that since then XDR-TB has spread from South Africa to Kenya, Tanzania, and Uganda. This raises the question of whether it might not be best to place XDR-TB patients under quarantine. And this is precisely what the US Centers for Disease Control and Prevention ordered at the end of May 2007 in the case of a patient believed to have XDR-TB. This was the first such isolation order to be issued in the USA since a case of smallpox in 1963.

The return of tuberculosis: a result of missed opportunities

At the end of the 20th century tuberculosis reappeared in Russia and many countries of the former Soviet Union. At that time a case rate of 85 tuberculosis patients per 100,000 inhabitants was reported in Russia. In Kazakhstan as many as 126 tuberculosis cases were reported per 100,000 inhabitants, in Kirghizia 123, in Georgia 96, and in Turkmenistan 90. Even Latvia, now a member of the European Union, reported a case rate of more than 80 per 100,000 inhabitants.

At the same time, however, the disease has also flared up in recent times in problem districts of the cities of Western industrialized countries. This was the case, for example, in central Harlem, a neighborhood of New York City, where in 1990 as many as 230 people per 100,000 became ill with tuberculosis. Two years later almost 4000 New Yorkers were reported to have developed tuberculosis. Thereafter the incidence fell – to about 1500 cases in 2000. But precisely at this time the problem became worse in London, UK. In 2000 about 4000 people there were reported to have developed tuberculosis. Overcrowded apartments, the demolition of whole terraces of houses, the increasing number of homeless people, rising immigration, urban squalor, and not least the increase in the number of HIV-infected people are the principal reasons for this resurgence of the disease in London. And precisely for these reasons, it is believed that a large number of cases go unreported.

What went wrong?

The causative organism has been known for over 125 years. Diagnostic tests, treatment, and a vaccine to prevent the disease are available. Why, then, is tuberculosis still such a problem today? The shameful truth is that in industrialized countries the problem has long been regarded as having been to all intents and purposes solved. Here, it is said, the disease has been largely overcome. In other parts of the world the situation has always been different: particularly in Africa, Southeast Asia, and the Pacific, the problem has never been solved. Yet the false sense of security felt in the West led to almost complete cessation of research and development activity. The result: all quiet on the Western Front. Tuberculosis is still diagnosed just as it was by Robert Koch over 125 years ago, namely by the demonstration of acid-fast bacilli in sputum. The tuberculin test is also more than 100 years old. The BCG vaccine was developed in the first two decades of the 20th century. And most antituberculous drugs first became available between 1945 and 1970. What followed was radio silence. Of the roughly 1400 new drugs that were approved for use in the final quarter of the 20th century, only three were intended for use in the treatment of tuberculosis.

We are now being punished for this complacency. Demonstration of acid-fast bacilli is not a sufficiently sensitive diagnostic method in many cases of tuberculosis. The BCG vaccine protects young children, but not their parents. And more and more tubercle bacilli are becoming resistant to the available drugs. The problem of tuberculosis is thus not least a problem of missed opportunities. HIV is now providing a huge impetus to the spread of tuberculosis in many countries, especially in Africa. And no end is in sight.

Unfortunately, tuberculosis is often underestimated precisely because of its tenacity and for this reason it often falls through the net of aid organizations. In 2005 the World Bank spent only 3.5 million dollars on the fight against tuberculosis, whereas 1.2 billion dollars went to HIV/AIDS and 167 million dollars to malaria programs in Africa.

5.6 AIDS and tuberculosis: two diseases, one patient

Worldwide, approximately 15 million people are infected both with HIV and with the tubercle bacillus. And each year two million more people with both infections are added to this list. In South Africa alone there are already two million such people. Fig. 13 shows the proportion of tuberculosis patients who are also infected with HIV. Tuberculosis is one of the commonest causes of death in HIV-infected people, especially in Africa. And it is this association with HIV that is causing the number of cases of tuberculosis to rise continuously.

Why is the combination of AIDS and tuberculosis so disastrous? Because these two pathogens complement each other in such an insidious way that the resulting combined disease is worse than the sum of the two individual diseases. A third of the world's population is infected with the tubercle bacillus. This microbe has lived in human beings for thousands of years and over this time has arrived at an apparently peaceful coexistence with the host in which up to 90% of infected individuals remain healthy. Yet the microbe persists – actively kept in check by the immune system – in the body of the host for the duration of that person's life. Any weakening of immunity can therefore precipitate a bout of tuberculosis, in most cases before any opportunistic pathogen can gain a foothold. By knocking out the control center of the immune system (CD4 T helper cells), HIV therefore makes the tubercle bacillus far more dangerous to its host. Coinfection with HIV increases the likelihood that clinical tuberculosis will develop by a factor of more than one hundred. It substantially reduces the latency period, i.e. the time between infection with the mycobacterium and the development of tuberculosis. In most cases clinical tuberculosis appears within a year of HIV infection. The situation is further complicated by the fact that tuberculosis generally shows an atypical course in AIDS patients, e.g. patients are often infectious even before their tuberculosis is diagnosed. This considerably increases the risk of contagion.

Infection with HIV is, at least in theory, avoidable. By contrast, there is little that one can do to avoid infection with the tubercle bacillus. It is scarcely possible to protect oneself against droplets that a coughing person spreads in the same room, bus, train, or airplane. The two microbes add to each other's effect and compound each other's danger. Now that they have become well adapted to each other, even an effective method of controlling AIDS would by itself no longer be able to solve the problem of tuberculosis. HIV has greatly increased the spread of tuberculosis. Many people are now at risk of acquiring the disease by respiratory infection.

Doctors at a loss
The choice of treatment regimen poses a particular problem. Every fifth HIV-positive person reacts to ART with an inflammatory syndrome due to recovery of the immune system. This causes tuberculosis to become more severe than it was before treatment. It may be that the returning CD4 T helper cells bring about a cytokine storm that results in exacerbation.

There is another way in which HIV promotes the spread of tuberculosis. In many African countries HIV-positive babies do not receive ART. Instead, ART is withheld from young children until they show evidence of AIDS. This is far too late. BCG immunization is also potentially problematic in that, since BCG is a live vaccine, it may itself cause illness in immunocompromised individuals. Though all available data suggest that this happens only rarely, the World Health Organization advises against BCG immunization of HIV-positive neonates. These

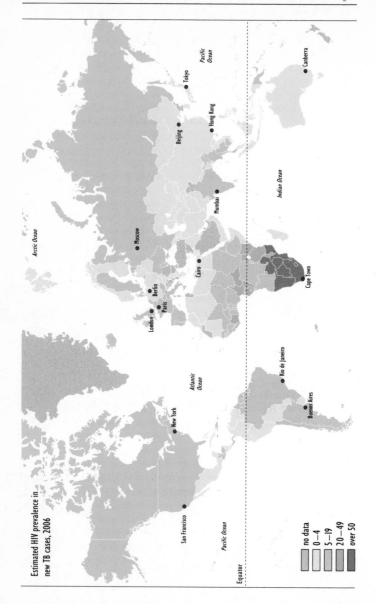

Estimated HIV prevalence in
new TB cases, 2006

Arctic Ocean

Pacific Ocean

Atlantic Ocean

Indian Ocean

Pacific Ocean

Equator

San Francisco

New York

London
Paris
Berlin

Moscow

Cairo

Beijing

Tokyo

Hong Kong

Mumbai

Cape Town

Rio de Janeiro

Buenos Aires

Canberra

no data
0—4
5—19
20—49
over 50

infants are therefore not only at risk of developing AIDS sooner or later, but also – because of their defective immune response – more liable to develop tuberculosis. A vicious circle has thus been created.

In general, HIV is likely to make immunization campaigns ever more difficult. This applies in particular to the immunization of neonates with live vaccines against measles, mumps, rubella, and poliomyelitis. In immunocompromised individuals every live vaccine has the potential to multiply unchecked and thus to cause illness. This is true of all immunizations in HIV-positive individuals. The weakening of the CD4 T helper cells, which play a central role in the generation of a protective immune response in all diseases, results in the immune response being weaker or completely absent. Even established immunity can collapse after HIV infection.

This is also a problem in relation to the development of new vaccines, e.g. against malaria, tuberculosis, and HIV/AIDS itself. In all likelihood, none of the subunit vaccines against tuberculosis that are presently undergoing clinical trials will prove to be any more effective than BCG. They are intended for booster immunizations that build on existing BCG-induced immunity. What happens, however, if this immunity collapses or is absent in the first place because an HIV-positive child has not been immunized? In such cases the possibility that subunit vaccines will build up a satisfactory degree of immunity against tuberculosis will also disappear.

One possible solution to this problem could be provided by a

Figure 13 Estimated proportion of people with tuberculosis who are also infected with HIV in the various regions of the world in 2006

Percentage of patients with tuberculosis who are also infected with HIV

Source: World Health Organization (2008)

new generation of genetically modified BCG vaccines. As well as stimulating a more powerful immune response than does BCG, these are safer to administer to HIV-positive neonates. The basic objective here is to alter the BCG vaccine in such a way that it dies at some stage after being administered and at the latest before an HIV-infected infant develops AIDS. At present, however, developments of this type are still at an early stage. At least ten years are sure to pass before they can be offered to those in need of them.

5.7 Malaria

A few fateful seconds

With its tiny stiletto it punctures the skin and inserts its proboscis to about half its length. If the proboscis enters a blood vessel, the mosquito injects its saliva through a tube. This relieves pain and prevents the blood from clotting. The mosquito sucks up the warm blood. A pinprick, a lump, a suggestion of an itch – that's all there is to a mosquito bite at European latitudes. Yet for the approximately 40% of the world's population that lives in malarious areas every mosquito bite can bring serious illness. Each year up to 600 million people develop this disease of the blood, and each year 1–1.5 million people – mostly in sub-Saharan Africa – die as a result of infection with these protozoa. Worst affected are children and expectant mothers. Each year 800,000 children die of malaria, a death rate of one child every thirty seconds. Fig. 14 shows the number of cases of malaria in the various regions of the world. Many other figures could be mentioned. For example, malaria imposes a huge economic burden on regions that are already poor. Thus, it is estimated that in sub-Saharan Africa the direct and indirect costs of the disease exceed two billion US dollars annually. In the USA about 1200 malaria cases are reported annually, mostly in travelers and immigrants from areas in which malaria is endemic.

Intermittent fever

The disease is detected primarily with the aid of a microscope, the diagnosis being made by the finding of parasites in the blood. Malaria causes bouts of high fever alternating with chills and is thus a form of intermittent fever. Joint pains, headache, stomach cramps, and vomiting may also occur. Severe malaria can cause renal failure, anemia, and jaundice and generally ends in coma. The parasites destroy the red blood cells (erythrocytes) and break down hemoglobin, the red pigment that transports oxygen in the blood. In this way they can substantially impair oxygen supply to the tissues. Malaria is not a one-off event: in endemic areas people can be infected a number of times over the course of their life. As people grow to become adults they gradually develop a degree of immunity. This explains why children are far more susceptible to the disease than are adults.

A flight of protozoa

Malaria is caused by a genus of protozoa known as *Plasmodium*. Four species of plasmodia are known to cause malaria, namely *Plasmodium falciparum*, which causes malignant tertian malaria, *Plasmodium malariae*, which causes quartan malaria, *Plasmodium vivax*, which causes benign tertian malaria, and *Plasmodium ovale*, which causes ovale tertian malaria. By far the most dangerous of these parasites is *Plasmodium falciparum*.

The protozoa arrive by air, transported by female anopheles mosquitoes. The parasites need the mosquitoes not only for transport, however, but also for their development. Of the approximately 400 different types of anopheles mosquitoes that

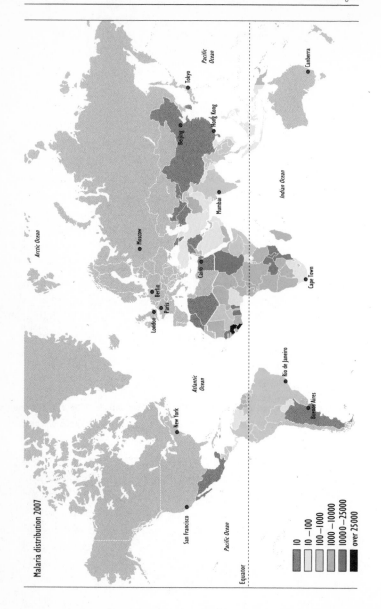

Malaria distribution 2007

10
10 – 100
100 – 1000
1000 – 10000
10000 – 25000
over 25000

are known to exist, sixty have the potential to act as malaria vectors. Insects are vectors of many other infectious diseases (see box: "Piggybacked into the blood").

Plasmodium has a complex life cycle both in the anopheles mosquito and in the human host. The radical changes that it undergoes during this life cycle make it very difficult to combat. On taking a blood meal, an infected mosquito transmits a form of the parasite known as a sporozoite to the human host. This intermediate stage of the parasite has previously migrated to the salivary gland of the mosquito. The insect injects the parasites into the human host along with its saliva. The sporozoites are then transported in the blood to the host's liver. There they mature and reproduce. Over the first two to three weeks after being infected the host experiences scarcely any symptoms. During this period merozoites are being formed. These then pass into the blood. From there they enter red blood cells, where they establish themselves, reproduce, and break down hemoglobin. In a similar way to certain viruses, the parasites convert the erythrocytes into a self-destructive factory. The erythrocytes duly disintegrate, thereby releasing merozoites into the bloodstream. These then enter more erythrocytes and in the process release toxins that initiate a cytokine storm. Abrupt fever, sweating, chills, and anemia are the result. In addition, the disintegrated erythrocytes damage fine blood capillaries, in particular the highly sensitive capillaries of the brain. This results in cerebral malaria, which often ends in the death of the patient. As in Spanish influenza and septic shock, overproduction of cytokines plays a role both in the fever and in the brain damage that occurs.

While all of this has been happening, some parasites have

Figure 14 Malaria distribution in the various regions of the world in 2007
Source: World Health Organization (2008)

entered the sexual stage of their life cycle, and it is in this stage that they find their way back into mosquitoes when these bite the patient. Thus the vicious cycle can begin anew. When male and female parasites meet in the gut of the anopheles mosquito, they produce offspring. After a phase of several weeks in the mosquito, the newly formed parasites are ready for their next attack on a human.

The myth of tonic water

In itself, malaria is curable if diagnosed sufficiently early and then treated immediately. The principal groups of drugs available are as follows:

Quinine, which has been used for medicinal purposes for over 400 years. As far back as the 15th century the bark of the cinchona tree was used in Peru to treat fever, however as quinine causes substantial side effects it is now used only in severe cases of malaria. Quinine-containing beverages such as tonic water are ineffective, since they contain less than 100 mg of quinine per liter, whereas a dose well in excess of 1,000 mg is required for effective treatment of malaria.

Chloroquine was introduced into malaria treatment in the 1950s. By the end of the 1950s resistant parasites had appeared, and by the end of the 1970s chloroquine-resistant parasites were present in most parts of the world. Use of the drug therefore had to be abandoned in most countries. A number of alternatives to chloroquine are now available, including *mefloquine*, *doxycycline*, and *atovaquone-proguanil*, however mefloquine resistance has now become common in some countries.

Artemisinin-containing preparations have been increasingly used in recent times. Artemisinin is a Chinese herbal product obtained from the plant *Artemisia annua*. Chinese people have long known that this type of wormwood plant can reduce fever.

Piggybacked into the blood

The secret of the success of many viruses is to be found in four, six, or eight legs and sometimes a pair of wings: the viruses have themselves transported from host to host by tiny animals. Without these vectors the viruses concerned would be vastly less dangerous and probably a lot less common. Certainly, European latitudes are not exactly the best place in the world to find such pathogens. Yet we have bugs, mites, fleas, lice, ticks, stinging flies, gnats, horseflies, and blackflies – and these are just a few of the potential bloodsucking vectors that we encounter here. Just one sting or bite and we can become infected.

Ticks are particularly feared in Europe. They can infect humans with the viral disease Russian spring-summer encephalitis, or RSSE, also known as tick-borne encephalitis, and also with borreliosis, a bacterial disease. In Germany alone several hundred people contract RSSE each year and about 60,000 contract borreliosis. RSSE starts with headache and limb pains and in about a tenth of patients attacks the meninges and in some cases the brain. Some cases are fatal. The onset of borreliosis is often announced by reddening of the skin accompanied by general malaise. In severe cases cranial nerve deficits also occur. A feared complication is a chronic inflammation of the joints known as Lyme arthritis.

Worldwide, a significant proportion of pathogens require the assistance of vectors. The most famous example is malaria, however dengue, yellow fever, West Nile encephalitis, and the chikungunya virus are also transmitted by mosquitoes. Virologists have coined a special name for this group of viruses, namely "arboviruses", an abbreviation of "arthropod-borne viruses".

It is cheap and effective: three days of treatment is enough to drastically suppress the parasite and reduce fever. Combination therapy based on artemisinin is currently the most economical and effective treatment available for the disease. Pediatric combinations of this kind include the combination of artesunate and amodiaquine. One pharmaceutical company has announced its intention of producing this combination without patent protection so that other manufacturers can also produce it. Instead of two different tablets that are intended for use in adults and consequently have to be divided in half for use in children, a single tablet containing both drugs is now to be produced. Use of a combination of two drugs increases the effectiveness of treatment and may reduce the likelihood that resistance will develop.

Protection during sleep

Since mosquitoes bite mostly at night, the best and cheapest method of controlling malaria has long been known to be mosquito nets, preferably impregnated with insecticides. Such nets are nontoxic to humans, whereas they repel mosquitoes or else kill them if they land on the net. Insecticide-impregnated bed nets whose effect persists for up to five years are now available. Previously, such nets had to be resprayed every six months. At present, however, only 1 to 2% of people who live in malarious areas sleep under mosquito nets.

Other preventive measures include anti-insect sprays for use inside houses. Insecticides are also used in this way. The best known, but also most notorious, of these is DDT, which was used successfully for controlling malaria in Southern Europe in the 1940s. Italy, Portugal, Spain, Bulgaria, Romania, Yugoslavia, and Hungary were rendered malaria-free in this way. The success achieved in those countries was also due partly to drainage of marshland. However, from the 1940s through to the 1960s DDT was used so extensively and in such uncontrolled fashion that environmental damage soon became apparent. We now know that when used sensibly and with suitable restrictions, DDT has a definite place in malaria control. The World Health Organization has therefore once again recommended the use of DDT for indoor spraying of houses (see box: "DDT and malaria control: not simply good or bad").

No single vaccine

There have been many attempts to develop a vaccine against malaria, and in principle immunization against malaria is possible. For example, it has been shown that heavily irradiated plasmodia can provide complete protection against malaria. Unfortunately, however, it has not proved technically possible to obtain the gigantic amounts of irradiated plasmodia that would be required for this purpose. Other approaches are therefore required. The particularly tricky problem with malaria is that a different immune response is required for each stage in the life cycle of the parasite. Whether a single vaccine can provide effective malaria control therefore remains to be seen.

Recently a small glimmer of hope appeared: a clinical trial performed in Mozambique with a vaccine developed by the pharmaceutical company GlaxoSmithKline showed a transient reduction of one-third in the number of cases of malaria in young children. In a study performed in The Gambia, use of the same vaccine also reduced the number of cases of malaria in adults. Though these figures may seem insignificant compared to those obtained with the vaccines used for other diseases, they represent considerable success in the case of malaria. The vaccine concerned works against the early stage of the malaria parasite in the human body. Preliminary trials of other vaccine candidates

DDT and malaria control: not simply good or bad

"My fly cage was soon so poisonous that even after intensive
cleaning of the cage, untreated flies fell to the floor as soon as they
came into contact with the wall. I was unable to continue until
I had changed the walls of the cage." This was how Paul Her-
rmann Müller described the experiments he performed in 1939 in
which he discovered the insecticidal action of DDT (dichlorodiphenyltrichloroethane). DDT had been synthesized as long ago
as 1874, however its potential for use in the control of malaria
was not recognized until much later. During the Second World
War DDT was used extensively. Over the ensuing period of thirty
years almost 500 million kilograms of it were sprayed around,
however claims that it was an environmental poison became ever
louder until in 1972 it was banned.

It soon became clear that the baby had been thrown out with
the bathwater. Though there is no doubt that the use of DDT
in agriculture caused a great deal of environmental damage, the
value of the substance in malaria control also needs to be recognized. It has now been agreed that DDT may be used for malaria
vector control in accordance with World Health Organization
recommendations and guidelines when locally safe, effective, and
affordable alternatives are not available. The World Health Organization recommends that DDT be used only indoors. Countries
are required to register for this purpose and report the amount
they have used every three years.

There is good evidence that when correctly used, DDT can
reduce malaria transmission by up to 90%. The mosquitoes
absorb a fatal dose as soon as they land on a wall that has been
sprayed. In India correct use of DDT has brought success, and
South Africa has now reintroduced DDT. At present indoor
residual spraying with insecticides is being practiced in fourteen
countries, ten of which are using DDT. If we are to achieve the
Millennium Development Goal of halting the spread of malaria,
we will not be able to get by without DDT – simply for lack of
more effective alternatives.

have also yielded promising results. Use of a combination of vaccines could perhaps enhance the effect.

5.8 Influenza in humans and birds

In early 2006 Germany, as other European countries, found itself in a state of emergency: fear of influenza gripped the nation. People with no medical background suddenly started using words like "pandemic" without batting an eyelid. Worried people ordered Tamiflu over the internet, bypassing doctors and pharmacists, and even virus-proof face masks were in demand. Specialist internet sites advised city dwellers on matters of survival: how much water or how many sachets of powdered soup should I stash away for my family of three people? Then came summer, and the hysteria about avian influenza gave way to rapturous joy at the soccer World Cup. What had happened? A pathogenic microorganism – H5N1, an influenza virus – had thrown the world into turmoil. The virus concerned has probably existed for millions of years, however only in the relatively recent past had it become pathogenic to humans.

The ABC of influenza viruses

Influenza viruses were initially responsible for diseases of poultry, or to be more precise, for diarrhea and the like in waterfowl. It is not clear when humans were first infected. Nowadays, in any case, some influenza viruses cross back and forth between poultry, humans, and other farmed animals such as pigs. This development has been due in large part to the conditions under which poultry are raised in Asia, where new strains of the virus are now appearing regularly. In this regard the main focus of attention has been on influenza A viruses. In addition to these, we know of influenza B viruses, which mostly infect humans, and influenza C viruses, which infect humans and pigs and generally cause only mild illness. Influenza B and C viruses have adapted well to humans and in so doing have become much less dangerous and lost most of their ability to change, quite unlike influenza A viruses, to which most of the following discussion refers.

Influenza A and B viruses are named in accordance with a strict formula that results in names such as A / chicken / Kyoto / 3 / 2004 (H5N1). The first item in the name indicates whether the virus is an influenza A or B virus. This is followed by the animal species from which the virus was isolated and the place where the virus was discovered, in this case Kyoto in Japan. Next comes a number indicating the strain, then the year, and finally, in brackets, the antigenic structure. The two most important antigens, namely the H and N proteins of the viral envelope, are more precisely characterized by means of a figure. H stands for hemagglutinin, a protein that causes blood to clump. N stands for neuraminidase, an enzyme that cleaves neuraminic acids from sugar components on the cell surface. These two proteins mediate the attachment of the influenza virus to cells of the respiratory tract.

The H and N composition of a virus determines, among other things, the specificity of the virus for its host. In addition, the host's immune system recognizes the virus mostly on the basis of its H and N proteins. As the principal antigens of the virus, these two molecules thus play a crucial role in resistance to infection and in immunization.

Runny nose, cough, sore throat

Influenza is transmitted by droplets (coughing, runny nose, sneezing), however it is by no means the same as a common cold, or catarrh. Though influenza often starts with cold-like symptoms, these are soon overtaken by high fever. Inflammation of the airways leads to severe coughing and general malaise. Serious complications such as cardiovascular failure and heart failure can occur. Especially in young children, elderly people, and immunocompromised individuals, influenza can be fatal. By damaging the airways the virus often smooths the path for secondary infections, mostly with bacteria, that can progress to pneumonia. The most important bacteria in this regard are pneumococci, *Haemophilus influenzae*, and staphylococci.

Virus blockers

Viral influenza can be treated using drugs such as amantadine and rimantadine. These block reproduction of influenza A viruses by inhibiting certain ion channels in the viral envelope. This prevents the viruses from releasing their genome into the host cell. The viral life cycle is thus interrupted. As the protein on which they act is present in influenza A viruses but not in influenza B viruses, these substances are ineffective against influenza B viruses. H5N1 is already largely resistant to these drugs. Chinese farmers administered the drugs to their chickens via drinking water. This was a pity, as amantadine and rimantadine are cheap to produce in large amounts and can be stored for a long time.

In 1999 Tamiflu (oseltamivir), a drug that inhibits the neuraminidase activity of influenza viruses, was introduced. A second neuraminidase inhibitor is Relenza, which, however, has to be inhaled. Inhalation is more complex and expensive than oral administration, but results in a rapid response. Both these substances are expensive and have a short shelf life. And influenza viruses, including H5N1, are now starting to become resistant to them too.

Vaccination

Every fall people are advised to have themselves immunized against influenza. Those who do so need a new shot every year, because every year or two mutation leads to new variants of the virus that differ slightly from one another. Scientists refer to these minor changes as *antigenic drift*. The small differences are sufficient to deceive our immune system. Though our immune system is quite capable of remembering a particular variant, no memory is of any use if the next variant differs to such an extent that it simply cannot be adequately recognized. The protection afforded by immunization is thus generally not sufficient for the wave of influenza that occurs in the following year. Each year, therefore, the new strains of influenza virus that have arisen in Asia, where the influenza epidemic starts, are typed and a mixture suitable for producing a new vaccine is put together. Each year the vaccine differs somewhat from that of the previous year. For the 2006/2007 season, use of an *A / New Caledonia / 20 / 99 (H1N1)*-like strain, an *A / Wisconsin / 67 / 2005 (H3N2)*-like strain, and a *B / Malaysia / 2506 / 2004*-like strain was recommended. So far vaccine production has gone well, at least in Europe and North America, as the vaccine manufacturers had sufficient time to prepare the vaccines. Global production of influenza vaccines is concentrated in nine industrialized countries, and vaccine manufacturers are currently able to produce about 350 million doses of vaccine each year.

Beyond all measure

Because the symptoms of influenza are fairly nonspecific, epidemics that occurred in the distant past are difficult to attribute unequivocally to influenza. For the 20th century, however, more precise information is available. In the following discussion the different epidemiologic patterns of influenza infection, i.e.

epidemics and pandemics, are considered in detail. Spanish influenza is used as an example of an influenza pandemic. And finally, avian influenza H5N1 is discussed in detail.

Influenza epidemics are a familiar occurrence. Every fall they spread out across the land. And far more people die as a result of them than is generally supposed: worldwide, some two to five million people become severely ill with influenza each year and between 300,000 and 500,000 people die of the disease. In Germany around 8500 deaths were reported in the winter of 2002/2003, however the real number may have been far higher: up to 20,000 people probably died as a result of influenza during that influenza season. In the USA the Centers for Disease Control recorded 36,000 influenza cases during the same season. The usual diagnosis is pneumonia. It is difficult to assign the cause of death unequivocally to influenza. In many cases it is difficult to determine whether influenza viruses or pneumococci are the cause of death, not least because influenza viruses often open the way to pneumococcal infection.

The economic burden of influenza must also be enormous. In the USA influenza is believed to cause annual losses of up to 90 billion US dollars, with an estimated 25 million people becoming ill with the disease and more than 40,000 dying as a result of it. In Germany more than one million working days are lost to influenza and 10,000 to 20,000 people are hospitalized because of the disease.

Influenza pandemics are worldwide outbreaks and are far less common than epidemics. The 20th century saw three:

– 1918: *Spanish influenza*, with up to 50 million deaths;
– 1957: *Asian influenza*, with about two million deaths; and
– 1968: *Hong Kong influenza*, with about one million deaths.

The fundamental precondition for an influenza pandemic is a totally new virus, one that is completely unknown to the human immune system. Such a variant can arise as a result of exchange of genes between two different influenza viruses. This reassortment of the viral genome is known as *antigenic shift*. Such an exchange of genes can occur when different influenza viruses such as an avian influenza virus and a human influenza virus meet in a host cell. This can occur either in poultry, in humans, or in an intermediate host such as pigs. It is now clear that the virus of Asian influenza and that of Hong Kong influenza both arose in this way. Since the genes are separate but situated alongside each other within the virus, exchange of whole genes is a simple matter.

It is also possible, however, for an avian influenza virus to adapt itself gradually and silently to humans and to alter its host range in a succession of small steps, i.e. via a series of antigenic drifts. This is probably how Spanish influenza arose in 1918. On that occasion an influenza virus crossed directly to human beings.

Spanish influenza got going at the end of the First World War, which by itself had claimed 10 million human lives. It could even be said that the First World War paved the way for Spanish influenza: soldiers lived crammed together in barracks, the war had left many people very weak, and the movement of soldiers across continents helped the virus to spread. The world's attention was first drawn to the new form of influenza when it appeared in Spain at the end of May, 1918. As Spain had not been involved in the First World War, it was less inclined than other countries to censor horror stories and consequently allowed the report of an outbreak of infectious disease to be transmitted to the whole world by cable. This led to the name "Spanish flu", which gradually imposed itself. At first other names were also used.

In Germany, for example, people spoke of *Blitzkatarrh* (sudden catarrh), while in Cuba the term *trancazo* (a blow with a heavy stick) was used.

The disease spread around the world in three waves. It first revealed itself in the spring of 1918 as a highly contagious, but relatively mild, illness. It then receded for a while. Half a year later a second outbreak occurred. By that time the virus had adapted itself to human beings so well that it was not only highly infectious, but also extraordinarily lethal. By September 11, 1918 the disease had reached the USA and by October of the same year 20,000 US citizens were dying of it every week. Life expectancy fell precipitously to thirty-seven years. By early 1919 the disease was raging in many parts of the world. Little more than a year after it first appeared the disease had carried off 50 million human beings – about five times as many as had died in the First World War and more than have been killed by any other disease, any natural disaster, or any war either before or since in such a short time. Up to one billion people fell ill, about two-thirds of the world's population at that time.

In a period of twenty-five weeks Spanish influenza killed twice as many people as AIDS has killed in twenty-five years, and this despite the fact that its mortality rate was "only" 5%. Admittedly, many of the deaths were due to superinfection with bacteria and at least some of these deaths would not have occurred if antibiotics for use against the major pathogens of pneumonia, namely pneumococci, staphylococci, and *Haemophilus influenzae*, had been available at that time. On the other hand, some of these bacteria are now becoming increasingly resistant to antibiotics.

Hardest hit by the disease were young adults aged between twenty and thirty-five years, not least because soldiers were drawn mostly from that segment of the population. Yet even among women it was the young who were most affected: 70% of the women who died from the disease were less than thirty-five years old.

Whereas with 'normal' influenza only one patient in one thousand dies, Spanish influenza killed up to 5% of the people who fell ill with it. Why was it so much more devastating than other influenza pandemics and epidemics? Some clues to this are provided by studies published at the start of the present decade. Sequencing of the genome of the Spanish influenza virus revealed that the virus was derived from an avian influenza virus. The virus was then reconstructed in a laboratory and its virulence tested in animal experiments. From the results of these it was concluded that the virus of Spanish influenza probably caused not only the typical signs and symptoms of influenza, but also a cytokine storm. Such a storm activates the immune system beyond all measure, leading to the production of enormous amounts of mediators of inflammation that rapidly cause death – especially in young adults with a healthy immune system.

Fowl plague becomes bird flu

A wave of influenza that occurred in the present century caused such a stir that it led to the renaming of a disease: what had previously been known as "fowl plague" was now referred to in the media as "bird flu". The new (unofficial) term draws attention to the relationship that exists between influenza viruses that affect different species, whereas use of the word "plague" emphasizes the extent of the devastation that the disease often causes in battery chicken farms.

On the whole, the virus is still lethal only to birds, in particular waterfowl and chickens. Only in isolated cases have humans been infected: according to the World Health Organization a total of 385 people had fallen ill with H5N1 by the middle of June 2008, and of these 243 died (see Fig. 15). On the other hand, when it does succeed in entering the human body the virus causes serious and often fatal illness. So far, however, the virus generally fails to dock successfully at the cells of the human respiratory tract. So far, that is. Because as Spanish influenza demonstrated, that could change. The number of human cases is increasing. More people died from H5N1 infection in 2006 than in 2005. Though some cases of mild H5N1 illness may not have been diagnosed or reported, it must be assumed that mild or even asymptomatic illness is extremely rare. In other words, once the virus infects the cells of a human being it often kills its new host.

The wave of H5N1 infection started in China. Now it has not only spread throughout Asia, but also gained a foothold in the Middle East, Africa, and Europe. Like Asian influenza in 1957, Hong Kong influenza in 1968, and SARS, the animal pandemic started in the province of Guangdong in southern China. The H5N1 virus was first demonstrated there in farm geese in 1996. H5N1 is probably a cross betweeen an H5 goose virus and an N1 duck virus that later jumped to chickens. A number of variants of the virus now exist alongside one other. Isolated cases of infection have also been reported in a number of mammalian species – not only pigs, but also cats, tigers, leopards, and dogs, among other species.

As the US doctor Michael Greger writes in his book *Bird Flu*, H5N1 is "a virus of our own hatching." In China some 14 billion chickens, geese, and ducks live, or rather vegetate, in animal houses. Three-story animal keeping is common: battery-caged laying hens are positioned directly over pig cages which in turn are positioned over fish ponds. The pigs eat the chicken droppings and the pig excrement serves as a fertilizer for aquatic plants and algae and thus as food for the fish. Practical, yes. But a paradise for intestinal pathogens, a category that includes avian influenza viruses, especially as pigs are a potential incubator

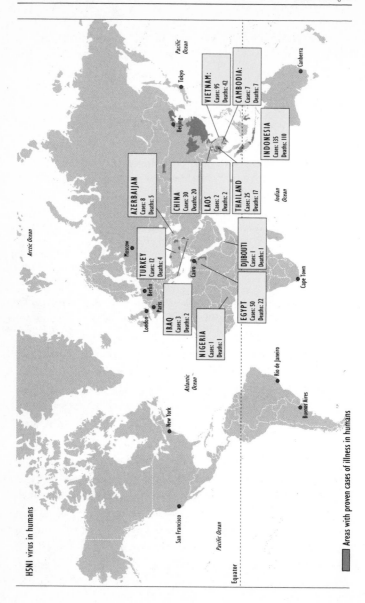

H5N1 virus in humans

VIETNAM:
Cases: 95
Deaths: 42

CAMBODIA:
Cases: 7
Deaths: 7

INDONESIA
Cases: 135
Deaths: 110

AZERBAIJAN
Cases: 8
Deaths: 5

CHINA
Cases: 30
Deaths: 20

LAOS
Cases: 2
Deaths: 2

THAILAND
Cases: 25
Deaths: 17

TURKEY
Cases: 12
Deaths: 4

DJIBOUTI
Cases: 1
Deaths: 1

IRAQ
Cases: 3
Deaths: 2

NIGERIA
Cases: 1
Deaths: 1

EGYPT
Cases: 50
Deaths: 22

Pacific Ocean
Tokyo
Beijing
Canberra
Indian Ocean
Arctic Ocean
Moscow
Cairo
Cape Town
Berlin
Paris
London
Rio de Janeiro
Buenos Aires
Atlantic Ocean
New York
San Francisco
Pacific Ocean
Equator

Areas with proven cases of illness in humans

for the development of new viral variants. From the breeding ground of battery farming the virus can move directly into the wild. There it can infect wild ducks and other birds. Yet these remain completely healthy and can distribute the virus over wide areas. Wild birds have thus become the Trojan Horse of influenza viruses.

Also dangerous are cockfights, in which blood spurts; H5N1 was probably transported from Thailand to Malaysia in this way. Until cockfighting was banned in Thailand because of H5N1, about 15 million cockfights were held every year.

Flu on the farm

However, H5N1 is not the only avian influenza virus that has made headlines in recent years. An outbreak of avian influenza due to the H7N7 strain occurred in 2003 in The Netherlands. The outbreak was ended by radical measures including the slaughter of 30 million chickens. Although about 1000 humans were infected and human-to-human transmission occurred, the virus proved to be relatively harmless to humans. One person died, a veterinarian who was directly involved in the culling. And new influenza viruses are appearing all the time in poultry farms. In most cases, however, the disease that they cause is referred to as fowl plague, its correct veterinary name.

Why, then, was there such a hysterical reaction to H5N1? For the reaction in many countries in the Western world was most certainly hysterical, even allowing for the fact that the potential risk posed by H5N1 is beyond dispute. The reason was simply the

Figure 15 Regions of the world in which people have fallen ill with H5N1 since 2003

Numbers of illnesses and deaths are shown
Source: World Health Organization (2008)

fear of another influenza pandemic on the scale of Spanish influenza. And when the facts are considered, this is a well-founded fear. In order to trigger an influenza pandemic, an influenza virus must satisfy three conditions:

1. It must cause influenza in humans.
2. It must reproduce in the human body.
3. It must be easily transmissible from human to human.

The H5N1 virus already satisfies the first and second conditions. The third condition could be satisfied either after an abrupt change or after a series of smaller changes. Here is a retrospective chronology of the events surrounding H5N1:

On May 14, 1997 in Hong Kong a young boy suddenly developed a high fever. A week later he was dead. The cause was found to be an influenza virus. And not an H1, H2, or H3 influenza virus, the types which until then had been thought to be the only ones that were pathogenic to humans. Instead, the fatal virus was H5N1. Until then, H5 had been known to be pathogenic only to poultry. Two months earlier it had caused an outbreak of disease in poultry in Hong Kong. Each day 100,000 chickens, ducks, and geese are transported – alive – to Hong Kong from Guangdong Province in China. In most cases they are killed, and killed without any special hygienic precautions, only when they reach the marketplace. Each year the eight million inhabitants of Hong Kong buy 38 million live chickens. There are millions upon millions of contacts between humans and these animals. In addition to the boy who was the first case, eighteen people fell ill with the H5N1 virus in 1997 and six of them died. In most cases the cause of death was multiple organ failure such as was also common in patients with Spanish influenza. Most of the people infected had either bought, sold, or processed chickens and had therefore been in close contact with poultry. Human-to-human transmission – the third precondition for a pandemic – was not found.

Active measures were soon taken in Hong Kong: more than one million chickens were slaughtered. The animals were burned to death, crushed to death, or buried alive. Yet the cause had not been eliminated. Given that China alone has a national poultry flock of 14 billion (more than twice the human population of the Earth!) and a national pig herd of 500 million, our chances of successfully eradicating H5N1 are now close to zero.

In 2001 there was another outbreak of H5N1 infection, once again in southern China. This time it was not until 2003 that the first deaths were reported in Hong Kong (see Fig. 15). Within a few months more than 100 million chickens in Asia had either died of H5N1 infection or been slaughtered in precaution-ary culling. After a brief respite the next outbreak occurred in 2004/2005. By June 2005 more than 100 cases of human illness and fifty-five human deaths had been reported.

By this time the virus had undergone considerable changes and at some stage a mutant variant had arisen. This was respon-sible for the outbreaks in 2004. As with the earlier variants, no human-to-human transmission was found with this mutant, however the virus had adapted itself perfectly to chickens and now pigs too were being infected. Fig. 16 shows the regions of the world in which H5N1 outbreaks have occurred in wild birds and poultry flocks.

In 2005 people's worst fears seemed to have been realized: the world reeled in shock at reports from Thailand of human-to-human transmission. An eleven-year-old girl who lived with her aunt had contracted the disease from infected chickens. She fell ill and died. Her mother, who had rushed back to the village from Bangkok to look after her daughter, also fell ill. And in all

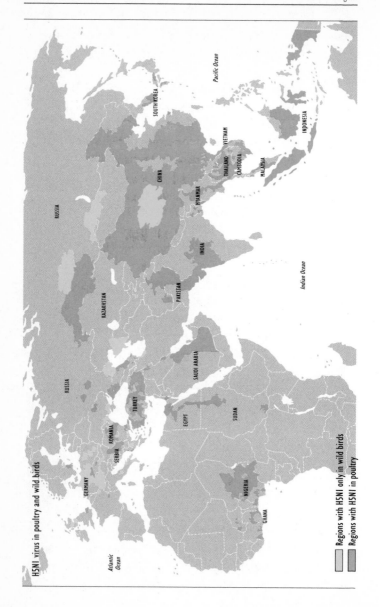

H5N1 virus in poultry and wild birds

Regions with H5N1 only in wild birds
Regions with H5N1 in poultry

probability she had never been in contact with infected chickens. She died soon after her daughter was buried. She appeared to have contracted the infection from her daughter. The girl's aunt also fell ill, but survived. Fortunately, no further cases of human-to-human transmission occurred. Close contact between the mother and her daughter had probably been the crucial factor.

In December 2006 a sixteen-year-old girl and her twenty-seven-year-old uncle died of H5N1 infection in Egypt. The virus concerned was found to have developed a degree of resistance to Tamiflu. Another milestone had been passed.

Treatment: In principle, the presently available anti-influenza drugs amantadine, rimantadine, Tamiflu, and Relenza also work against H5N1. However, some limitations have become apparent. For one thing, the virus is already resistant to amantadine and rimantadine, therefore treatment with these drugs is not a possibility. There have also been reports of Tamiflu-resistant H5N1 viruses. A number of industrialized countries have stockpiled Tamiflu for use in an emergency, however in developing countries and even in rapidly industrializing countries such a measure would swallow up the entire healthcare budget. It must be assumed that no country is adequately prepared for an H5N1 pandemic – not even the USA, which has stockpiled large amounts of Tamiflu and to date is probably the country that has reacted most sensitively to the threat of possible epidemics. Efforts to increase vaccine production have also been hugely intensified in the USA.

Figure 16 Regions of the world in which H5N1 outbreaks among birds have been reported

The illustration shows regions in which only wild birds have become ill and regions in which both poultry flocks and wild birds have been affected
Source: World Health Organization

Immunization: In principle, it should be possible to develop a vaccine against a potentially pandemic form of influenza such as H5N1. However, the technologies that are available at present are all oriented towards producing specific influenza vaccines, that is to say vaccines that act specifically against an existing viral variant. This requires precise knowledge of the structure of the virus. This means that it will not be possible to develop a vaccine against an H5N1 variant that becomes infectious to humans until precisely that strain has been identified. And even then a period of at least six months of development is required before production of a "classical" specific vaccine can commence. It is therefore to be feared that the vaccine would come too late to prevent the first wave of influenza. Nor would the number of doses that could be produced be sufficient. At the most optimistic estimate and assuming the use of a few manufacturing tricks, no more than one billion doses could be produced. Which segments of the population would then be preferentially immunized is difficult to predict. What's more, the vaccine industry is concentrated in a small number of countries: the USA, Canada, Australia, Japan, and a number of European countries have the capacity to produce vaccines, whereas no such capacity exists in many parts of Asia or anywhere in Africa or Latin America.

In the medium term research may provide solutions, since we have the know-how to develop vaccines against a broad range of pathogens, i.e. vaccines that can protect against a number of influenza variants. This could be extremely helpful not only for controlling avian influenza but also for preventing the annual influenza epidemic. Unfortunately, this strategy has been adopted very late, though hopefully not too late. From the point of view of vaccine manufacturers it is much more economically attractive to produce and sell a tailor-made, highly specific vaccine every year than to produce broad-spectrum vaccines for booster immunization every ten years. The first vaccine for use against H5N1 was recently approved in the USA – though not for trade, only for emergency storage. Unfortunately, the vaccine concerned is only about 50% effective and must therefore be regarded as only an interim solution.

And what is the situation regarding vaccines for poultry? They exist, but most of them protect only against illness, not against infection. This is sufficient for human immunization, however in veterinary medicine we need vaccines that also prevent infection. This is because infected birds transport the virus and excrete it, and if they don't become ill this goes unnoticed, as it does in many wild waterfowl. In China mass immunization of poultry is even said to have been responsible for the fact that an aggressive H5N1 strain spread and displaced other strains. According to US researchers, the vaccine was probably ineffective against precisely that strain – and thus contributed to its selection.

What now?

There has been much speculation as to how rapidly H5N1 could spread across the world if it developed the ability to pass directly from human to human. SARS has shown us, better than any computer model could, just how suddenly such a time bomb can explode. A single person is sufficient to initiate a process that can reach anywhere in the world within twenty-four hours. In the case of SARS things got better, whereas in the case of H5N1 the chances are not so good. In the case of SARS people fell ill before they became infectious, whereas in the case of influenza people are already infectious before they fall ill.

One thing is clear: it is now too late to eliminate H5N1 from its reservoir. We must learn to live with the virus. We therefore need to think about how to bring an H5N1 pandemic under control as rapidly as possible. This will depend upon our ability to interrupt transmission of the virus as soon and as systematically as possible. The first step is to limit the spread of the disease by giving preventive treatment with Tamiflu, i.e. by providing Tamiflu cover. A precondition for success here is that the outbreak be discovered promptly and that infected people be treated immediately – before the virus can be transmitted to people in contact with them.

Whereas many industrialized countries have stockpiled Tamiflu, most countries in Asia have not. It is precisely in Asia, however, that a pandemic virus is most likely to arise. Whether the industrialized countries will be prepared to hand over their stocks of Tamiflu at the start of an H5N1 pandemic remains to be seen.

Also necessary are strict quarantine measures. Here again, however, doubts as to the likelihood of success are in order. In the case of SARS it took months before the responsible authorities issued any pronouncements. No countermeasures were taken until hundreds of people had been infected. China has already done too much covering up in relation to H5N1 for us to feel confident that it would take prompt and appropriate countermeasures.

How hard an influenza pandemic would hit us is difficult to say. Nevertheless, the likelihood that a variant of the H5N1 virus that has adapted itself to humans will spread rapidly across the world is high. The consequences – of which SARS provided just a foretaste – for the world economy would also be dramatic. Estimates of the possible number of deaths are also in circulation, however some of these need to be treated with caution. Taking Hong Kong influenza as the mildest, and Spanish influenza as the deadliest, influenza pandemic of the twentieth century, an H5N1 pandemic could cause between eight million and 200 million deaths. Very pessimistic estimates go up to 1.5 billion deaths.

5.9   SARS: half way around the world in twelve hours

New plagues can appear at any time and anywhere. That's the theory. And recently the whole world saw precisely what this means in practice. The new millennium was barely two years old when SARS (severe acute respiratory syndrome) appeared. Out of the blue a large number of people fell ill with a mysterious form of pneumonia accompanied by high fever. And while scientists worked flat out, but for the time being unsuccessfully, to find the pathogen, the pathogen took off at unheard-of speed on a trip around the world. In markets, in hotels, and above all in hospitals it jumped from one person to another and traveled inside its human hosts on airplane trips between continents. We saw how a single ill person passed the infection on to a group of healthy people to form a cluster that then grew bigger. And we witnessed how the virus was exported out of this cluster to give rise to completely new clusters at places far removed. In short, this was a stunning example of the spread of contagious disease.

A hotel, a hospital, an airplane

On November 16, 2002 a forty-five-year-old man from Guangdong Province in southern China fell ill with fever and pneumonia. Four of his relatives probably caught the infection. The cause was unknown. On December 10, 2002 a cook working in a restaurant in Shenzhen developed similar symptoms independently of the forty-five-year-old. The cook was admitted to his local hospital, where he infected eight more people. All became ill. By the end of January 2003 the number of such cases in China had risen dramatically. Despite attempts to cover up the outbreak, news gradually filtered through to the public. The US Centers for Disease Control (CDC) sent a group of experts to southern

China. In early February 2003 the CDC experts started work on site. Eventually the news that a new infectious disease had been discovered reached the World Health Organization and other international bodies. But by then the responsible organism had moved one step ahead by finding its first 'super-spreader'. Over a period of about three weeks after being admitted to hospital in Guangdong, an ill fish merchant infected nineteen relatives and at least fifty doctors, nurses, and other health workers. Among these people was a sixty-four-year-old doctor. This doctor had probably had no direct contact with the fish merchant, but rather had acquired the infection from some other person working in the hospital. In mid-February he became ill, but not for long. On February 21, as he was feeling better, he and his wife made the three-hour trip to Hong Kong to attend a wedding. There the couple booked themselves into the Metropole Hotel. Very soon the sixty-four-year-old doctor began to feel ill again. So ill, in fact, that on the following day he presented to Kwong Wah Hospital and was duly admitted. There he died on March 4. Yet the single night that he spent on the ninth floor of the hotel had been sufficient to infect at least sixteen other hotel guests and a hotel visitor. It is still not entirely clear how this happened. All the infected people probably used the same elevator and they may all have pressed the same elevator button. Nevertheless, it remains a mystery why more people did not become ill and in particular why none of the hotel porters, chambermaids, receptionists, or cleaners became ill. In any case, the four-star hotel became the point of departure for the (as yet unknown) virus's trip around the world.

A forty-eight-year-old businessman from New York checked out of the hotel in Hong Kong and flew on to Hanoi, Vietnam. There he infected sixty-three more people. A woman carried a second load of the virus to Singapore, where it infected at least

195 more people. A seventy-eight-year-old Canadian woman who was infected with the virus flew to Toronto and there infected 136 people. Within a few hours this new infectious disease had spread from Asia to North America. Meanwhile in Hong Kong, another Canadian became so ill that he took himself to the Prince of Wales Hospital. There the virus spread to nine other people. A twenty-six-year-old worker who had acquired SARS when he briefly visited the Metropole Hotel on that fateful night was admitted to the same hospital. On becoming ill, this young man had at first thought nothing of it. On March 4 he was admitted to the Prince of Wales Hospital. There he infected 143 people, including a thirty-three-year old dialysis patient who became infected at one of his dialysis sessions. From the hospital this dialysis patient returned to his apartment in a housing estate containing 19,000 residents. On March 14, 2003 he fell ill with pneumonia. He was admitted briefly to hospital but then discharged too early. He became ill again. A single night at home was enough for him to infect 213 other residents of the housing estate with the virus. The virus was probably transported from apartment to apartment via the sewage system. Also infected in the Prince of Wales Hospital was a sixty-two-year-old man from Beijing who was visiting his sick brother. The visitor soon fell ill, but refused to go to hospital. Instead, he flew home. On March 15 he took Flight CA 112 from Hong Kong to Beijing. Five days later he died. But during the flight the virus had spread to at least twenty-two other passengers and two members of the crew. The SARS virus then traveled with these people to Taipei, Bangkok, Singapore, and even Inner Mongolia.

Meanwhile the task of identifying the agent responsible for the disease had been commenced in Vietnam. On March 3 in Hanoi Dr. Carlo Urbani, of the World Health Organization, visited the businessman who had come from the Metropole Hotel in Hong

Kong. Dr. Urbani imposed strict safety measures and sent blood samples and throat swabs off to specialized laboratories including the US CDC. This was a decisive step forwards, since from that point on the public was made aware of, and became alarmed by, the new disease. In Vietnam the SARS virus was overcome relatively rapidly. However, Dr. Urbani, who had announced the need for a global fight against SARS, made one mistake: he himself became ill – either shortly before or during a flight from Hanoi to Bangkok, where he was intending to give a talk on tropical diseases. On March 29 the expert died in Thailand. Fortunately, he did not pass the virus on to anyone else either in the airplane or in Bangkok.

SARS traveled around the world with impressive speed (see Fig. 17). Yet things could have been a lot worse, for a number of the paths taken by the SARS virus turned out to be dead ends. For example, a married couple, both of whom were infected, traveled from the Metropole Hotel in Hong Kong to the Philippines. There they were admitted to hospital but soon recovered. Only later, when they went to Great Britain, did a test confirm that they had both survived SARS. They had not infected anyone. A similar sequence of events occurred in the case of a German woman who flew from Hong Kong to Australia, where she recovered spontaneously after a brief illness. Another man fell ill on his way to Vancouver, however on arrival there he was placed in complete isolation and treated. Once again, nobody else was infected. There was also a brief scare in Germany. On March 15 a doctor who was on his way back to Singapore after attending a conference in New York made a stopover in Frankfurt am Main. He had acquired the infection from patients in Singapore before traveling to the conference. Yet before flying out of New York the thirty-two-year-old doctor had told a colleague that he was experiencing SARS-like symptoms. This colleague then rang the German authorities from New York. The World Health Organization identified the flight. On arrival in Frankfurt the young doctor and his companions were immediately isolated and taken to hospital. The man was indeed suffering from SARS. Yet this virus too had failed: the doctor had infected his wife, his mother-in-law, and a flight attendant, but none of these people became ill. Nor did any of the conference participants or any of the other passengers on the flight develop the disease.

The subsequent evolution of the epidemic was as follows: on April 23 the number of people with SARS passed 4000; five days later the figure had climbed to 5000, by May 2 to 6000, and by May 8 to 7000. During this peak period more than 200 new cases were being reported each day. Then the epidemic subsided. Its cause had been known since late March. At first the much better known avian influenza virus H_5N_1 had been suspected, however it was then found that SARS was due to a new corona-virus that differed markedly from the coronaviruses previously known to infect humans and animals. Until then this family of viruses had been known mostly as a cause of harmless colds in humans. Shortly after the virus was identified a diagnostic test was developed. Cases of SARS kept appearing until the middle of 2003 and isolated reports of new cases continued through the second half of 2003 and into early 2004. Then the disease disappeared from the scene. More than 8000 people had developed the disease, and of these about 750 died.

The search for causes

Where had SARS come from? The precise details of the sequence of transmission have yet to be fully elucidated. Yet in all probability this was a typical zoonosis, i.e. a disease caused by a pathogen that jumped from animals to humans. It probably originated in meat markets in China. A high proportion of the first people

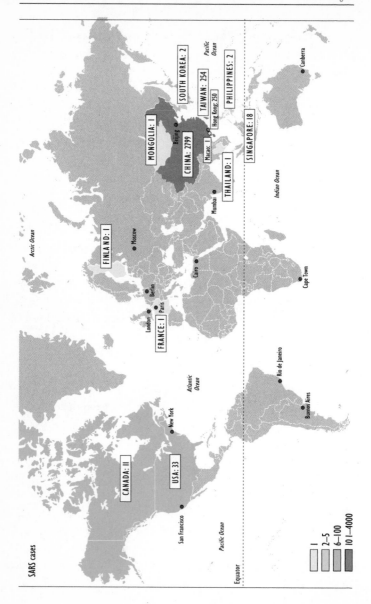

SARS cases

MONGOLIA: 1

SOUTH KOREA: 2

TAIWAN: 254

Hong Kong: 250

CHINA: 2799

Macao: 1

PHILIPPINES: 2

THAILAND: 1

SINGAPORE: 18

FINLAND: 1

FRANCE: 1

CANADA: 11

USA: 33

Beijing

Moscow

Berlin

London

Paris

Mumbai

Cairo

Cape Town

New York

San Francisco

Rio de Janeiro

Buenos Aires

Canberra

Pacific Ocean

Indian Ocean

Atlantic Ocean

Arctic Ocean

Pacific Ocean

Equator

1
2–5
6–100
101–4000

to be infected had either sold or processed wild animals for human consumption. Antibodies against the SARS virus were also found in the blood of many butchers and kitchen workers who were not ill, especially those who handled a species of viverrine known as the masked palm civet, a wild animal that is eaten as a delicacy in China. These animals, together with badgers and raccoon dogs, are now thought to be reservoirs of the virus. According to the best available evidence, the virus jumped from bats to civets and from there to humans. People who worked in meat markets were probably crucial for the change of host. It was in them that the SARS virus first developed into the form that is dangerous to humans. A single mutation was probably enough to make this happen: of the 1255 amino acids present in the adhesion molecule with which the virus attaches itself to body cells like a prickly bur, only one needed to be changed. Such mutations occur regularly in viruses. If a virus is in close contact with a potential new host, it can soon spread after undergoing such a change. The first market workers to become infected with the new, aggressive form of the virus probably had a degree of immunity to it as a result of previous contact with the original, relatively nonvirulent form. For this reason they did not become ill.

Ancient combat strategies in a globalized world

SARS killed almost 800 people and briefly paralyzed parts of the world economy. The outbreak cost Asia at least 25 billion dollars, while the international air travel industry sustained losses of six billion dollars, with Lufthansa alone losing 1.5

Figure 17 Number of people who have fallen ill with SARS in the various regions of the world since 2003

Source: World Health Organization

billion dollars. Yet compared to what could have happened, the world got off lightly. In the first place, the virus apparently failed to establish itself in humans on a permanent basis. Secondly, the illness weakened its victims too rapidly and too much for them to be able to transport it over long distances. Thirdly, those infected probably became infectious to others only after falling ill themselves rather than while they were still moving about and apparently healthy. And fourthly, the virus originated in a relatively small animal reservoir. Ultimately, therefore, SARS also served as a lesson on how an epidemic needs to be managed in the modern world. On the basis of the four characteristics of the virus referred to here, SARS was overcome using basically simple, centuries-old methods, namely quarantine, rigid travel restrictions, and slaughter of reservoir animals. The clinical thermometer, which was used briefly to measure the temperature of many air travelers, proved to be one of the most important tools in this respect. The slaughter of about 10,000 captured masked palm civets in Guangdong eliminated the source of the epidemic. These methods are far more difficult, or liable to fail altogether, in the case of pathogens that hide themselves away in niches or spread without giving any warning signals, as does HIV, for example. Also more successful are microbes that have access to an enormous animal reservoir, as for example do influenza viruses, which in China alone can hide themselves away in 14 billion chickens, geese, and ducks.

In the case of SARS, epidemiologists and health organizations were also able to exploit the possibilities offered by the 21st century, namely rapid transmission of information via the internet and the media and coordination of measures by international organizations, in particular the World Health Organization. It must also be noted, however, that modern technologies, in particular widespread rapid air travel, hugely facilitated the spread of SARS. Within half a day the virus crossed from Asia to Europe, Australia, and North America. And the media reporting of the epidemic caused panic in large sections of the population.

Despite the existence of the mass media and rapid transmission of information, however, it was some time before the public was told what was going on. For months China suppressed news of the emerging epidemic. Had this not been the case it might have been possible to stop the epidemic before it spread any further. What we urgently need for future epidemics – of SARS or of any other infectious disease – are better systems of surveillance and early detection that inform us of any new outbreak as rapidly and as transparently as possible. For nobody knows where the SARS virus has gone or whether it will return some day. In our global village attempts to seal ourselves off are doomed to failure. And a global approach also makes perfect economic sense. With the minimum of 25 billion dollars that SARS is estimated to have cost (other estimates range up to 100 billion dollars) most of the UN Millennium Development Goals for Africa could have been achieved.

5.10 Life in the shadowlands: the neglected tropical diseases

It's a perfect recipe for indifference: no explosive outbreaks, no risk to other continents, and a geographic distribution limited largely to remote rural areas of Africa, Asia, and Latin America. A disease that fits this description will never become the subject of a science-fiction bestseller. It will get scant attention in the media, and decision-making bodies will find it easy to look the other way. It will simply have to live in the shadowlands. And there are many such diseases: bilharziasis, elephantiasis, river blindness, sleeping sickness, Chagas disease, kala azar, and tropical ulcer, to name but a few. The World Health Organization lumps them together as "neglected tropical diseases". Most are caused by parasites and occur in desperately poor regions. There's no way of making money out of them, so they are largely ignored by the pharmaceutical industry and the world of science. The countries in which they occur don't have the means to deal with them. So it is that a few ancient diseases are still making life difficult for one billion people, i.e. a sixth of the world's population. Or, as stated by the World Health Organization, these diseases anchor one billion people and whole regions of the world in poverty. The neglected diseases form part of a vicious circle of poverty: a country in which a high proportion of the inhabitants are ill or permanently debilitated cannot make economic progress, and without economic progress there can be neither clean water nor sufficiently hygienic living conditions; instead, conditions will be perfect for the spread of precisely these diseases.

At a meeting on this subject convened by the World Health Organization in April 2007, the Vice President of Tanzania, Ali Mohamed Shein, stated: "I would like to emphasize that all these diseases are not neglected in any way by the developing countries. We in Tanzania, for example, have recognized since the dawn of independence that health status and health service delivery are the core of socioeconomic development. We have waged a protracted war against all diseases which are regarded as a hindrance to development alongside poverty and ignorance."

Drugs – many of which were originally developed for use against completely different pathogens of human beings or for use in farm animals – are available for use against some of the responsible parasites. However, more research and new developments are urgently required, especially as resistance to many drugs is increasing. Yet almost nothing is happening in the field of research and development. Only one dollar out of every

100,000 dollars spent on biomedical research goes to neglected tropical diseases. In response to this situation the World Health Organization has announced a new strategy of mass preventive chemotherapy with antihelmintics and antiparasitics such as albendazole, ivermectin, and praziquantel. This is because many of the people affected by these diseases have no access to medical care and because many of these diseases do not make their victim ill until long after infection occurs, with the result that not everybody is aware of the need for prevention. Compared to mass prevention of this kind, identification and curative treatment of individual cases are very difficult. A number of examples of neglected tropical diseases are discussed briefly below.

Protozoa on the march

Many people have heard of sleeping sickness. The causative agent of the disease, a trypanosome, is transmitted to humans by the bite of the tsetse fly in Africa. Trypanosomes are elongated protozoa with a whiplike tail known as a flagellum. They are about the size of a red blood cell. In late stages of the disease patients become emaciated and comatose. Coma often leads to death as the parasite attacks the nervous system. Autoimmune reactions with inflammation of the brain and heart form part of the clinical picture. About half a million people suffer from sleeping sickness and every tenth sufferer dies. A related pathogen, also a trypanosome, is responsible for Chagas disease in Latin America. In this case the vector is an assassin bug of the reduviid family. The most notable effects are on the cardiovascular system, here too due to autoimmune reactions. Here too the mortality rate among the approximately 200,000 people who fall ill with the disease each year is about one in ten. Drugs are available for both diseases, however their side effects are substantial and their effectiveness is limited. Unlike in malaria, mosquito nets are of little use for preventing sleeping sickness, since tsetse flies are active by day. As in the case of the assassin bugs, that leaves open the possibility of using insecticides.

Many neglected tropical diseases disfigure and disable their victims, who become socially excluded as a result. Leishmaniasis is an example of this. The flagellated protozoa that are responsible for this disease are found in Latin America, Africa, and India, and less commonly also in Europe. About 12 million people are infected. Transmission is by the bite of a sandfly. Three forms of leishmaniasis are distinguished. The first is visceral leishmaniasis, also known as kala azar or black fever. Patients with this form of the disease suffer irregular bouts of fever and their immune system fails; if left untreated, most die of opportunistic infections. The second form is cutaneous leishmaniasis. This consists of tropical ulcers and a group of related skin conditions. Near the sandfly bite there appear ulcers that later heal to leave in some cases enormous scars. In mucocutaneous leishmaniasis, the third form of the disease, the infection spreads especially in the mucous membrane of the nose and throat, causing overgrowth of tissue in the throat and destroying the nasal septum. Depictions of such disfigured faces are seen in artifacts dating back to pre-Inca times in Peru and Ecuador. As the depictions concerned date back to the 1st century AD, it is assumed that cutaneous leishmaniasis already existed at that time. A number of drugs, many of which are based on antimony, are now available for the treatment of leishmaniasis, however the results of treatment are not satisfactory. Resistance is increasing and side effects are serious. Another problem is that the vectors are so tiny that they pass through the mesh of standard mosquito nets. The best method of prevention is therefore to kill sandflies using insecticides.

The diarrheal disease amebic dysentery, which affects up to 50 million people worldwide each year, is also caused by protozoa. These microbes are spread via contaminated drinking water and foods and by fecal-oral transmission. As well as causing diarrhea, abdominal pain, and bloody feces, these amebas can spread throughout the body and affect internal organs, most commonly causing abscesses in the liver. Drugs are available, however the body does not develop lasting immunity, so people become ill again and again. Clean water and better sanitation would solve the problem.

Worms

Many tropical diseases are caused by worms, and the number of people affected is huge. Just to give some examples, 120 million people suffer from elephantiasis and similar diseases, 17 million from river blindness, and about 25 million from Calabar swelling. The causative agents of all these diseases are threadworms whose larvae are transmitted by insects. The agents responsible for elephantiasis – threadworms with the exotic names *Wuchereria* and *Brugia* – cause inflammation of lymphatic vessels that is often made worse by bacteria and fungi. In this way the threadworm disrupts the flow of lymph and can cause massive swelling of the legs. Hence the name "elephantiasis". *Onchocerca*, another threadworm, lives in the skin for more than ten years, there producing large numbers of larvae that cause inflammation in the skin and that can, among other things, find their way into the eyes and cause blindness. Hence the name "river blindness". Drugs are available for use against these diseases, however they are not optimal. This can be seen from the example of river blindness: ivermectin is effective at killing the larval stages of the parasite, but does not harm the female worms, which therefore keep producing new larvae. For this reason ivermectin has to be taken for a long time – until all the adult worms have died. Now that the drug has been used extensively in humans, some worms are developing resistance to it. Ivermectin was in fact originally developed for use in veterinary medicine.

Many people have heard of schistosomes, the type of worm that causes bilharziasis, or schistosomiasis. Certain developmental stages of these worms exist in water and bore their way through the skin of their hosts. In humans schistosomes live in blood vessels and consume red blood cells. Over 200 million people are infected with them. Their intermediate hosts are freshwater snails that are found in Africa, the Middle East, East Asia, and Latin America. In bilharziasis an initial phase of allergic inflammation of the skin is followed by bouts of fever and in many cases by acute involvement of other organs. Later the disease progresses to a third, chronic, phase brought about by the body's immune reaction against the eggs laid by the worms. Most of the eggs are carried in the blood to the liver, where they cause inflammation that ultimately leads to liver cirrhosis. Cancers of the bladder and urogenital tract can also arise. Praziquantel, a drug developed thirty years ago by German pharmaceutical companies in collaboration with the World Health Organization, is the drug of choice. At a recent conference in Geneva one of the giants of the pharmaceutical industry promised to supply 200 million doses of praziquantel at no cost as a contribution to achieving the Millennium Development Goals.

6 Antimicrobials

> You have to run as fast as you can just to stay where you are. If
> you want to get anywhere, you'll have to run much faster.
>
> Lewis Carroll

6.1 Antibiotics

Almost everyone has at some time come away from a doctor's
consultation with a prescription for an antibiotic. In 2005 alone
doctors in Germany – a country that is relatively restrained in its
antibiotic consumption – prescribed more than 270,000 doses.
By comparison, a staggering 40 million doses were prescribed in
the UK. In the USA doctors prescribe 150 million doses annually on top of the 190 million doses administered daily in US
hospitals. Probably as many as one half of these doses are inap-
propriate. The top position in Europe is held by France, where
more than 3% of the population takes an antibiotic every day. In
Germany, Austria, and the Netherlands only 1% of the population resort to antibiotics with such regularity. These drugs often
bring quick relief to patients with bacterial infections. They
specifically kill bacteria or at least slow their growth. Many
antibiotics are produced by bacteria or fungi as a sort of luxury
product, i.e. they are not essential for growth and reproduction.
However, for millions of years they have conferred a survival

advantage on those organisms that produce them. Other antibiotics serve as chemical signals and only incidentally have antibiotic effects. It was just a matter of someone discovering them.

The first to do so was Alexander Fleming (1898–1955), who discovered penicillin serendipitously. The substance is produced by the *Penicillium notatum* mold to defend itself against invasion by other microorganisms. Humans have also synthesized antibiotic substances, known as antimicrobials. Paul Ehrlich (1854–1915) was the father of the concept of antimicrobial therapy. He developed Salvarsan, the first effective drug for syphilis. Gerhard Domagk (1895–1965) later developed the sulfonamides, antimicrobials that are still in use today in modified form.

The terminological distinction between antibiotics and antimicrobials soon became academic when scientists learned how to modify naturally occurring antibiotics to enhance their efficacy or improve their tolerability. In the following I will therefore use the term "antibiotic" to denote any antibacterial substance and the term "antimicrobial therapy" to denote treatment with them. The term "anti-infective" refers to any substance capable of killing infectious agents or preventing them from causing infection and therefore also includes vaccines.

A primer on killing bacteria

The multitude of antibiotics now available have one thing in common: They fight bacteria by interfering with bacteria's unique metabolic processes, while ideally leaving the body's cells unharmed. Some, for example, disrupt the synthesis of the bacterial cell wall. They include the penicillins and cephalosporins. The members of this antibiotic group are called beta-lactams, so named after a basic chemical structure they all share. Penicillin attaches to proteins in the bacterial wall that usually form cross-links between molecules making up the wall. As a result, the wall structure becomes loose, and holes form. The porous bacterial cell then dissolves. However, penicillin cannot dispatch a bacterium at rest in this way. It works only when the cell wall is actively being synthesized, for example when the bacterium is growing and starts to divide. Vancomycin also inhibits cell-wall synthesis, but via another mechanism. Bacteria that are resistant to beta-lactam antibiotics may therefore still respond to treatment with vancomycin.

Other antibiotics act by disrupting bacterial protein synthesis. They include the aminoglycosides, the tetracyclines, chloramphenicol, and the macrolides. Even in cases where bacteria produce proteins that are similar to those in our own cells, the bacterial synthesis pathway still has many unique features. Also, because the synthesis process is similar in many bacterial strains, protein synthesis inhibitors are often able to kill several groups of bacteria at one fell swoop. Such agents are referred to as broad-spectrum antibiotics.

The quinolones are a new generation of antibiotics. They include ciprofloxacin, which made headlines after the anthrax attacks in late 2001. They act when the bacteria are multiplying. They kill bacteria by attaching themselves to an enzyme that uncoils DNA and splices cut pieces of DNA together again. Because the quinolones are relatively new, resistance to them is not so widespread. Other antimicrobials, such as the sulfonamides, or sulfa drugs, interfere with bacterial metabolic processes. The polymyxins act as detergents, dissolving the bacterial cell membrane. And because the bacterial cell membrane differs from that in the body's cells, the effect is specific.

Although the treatment of viral diseases is not as advanced as the treatment of bacterial diseases, significant progress has also been made in this area in recent years. However, the details would take us well beyond the scope of this book. The most important drugs for the treatment of influenza and AIDS are discussed in the relevant chapters. Several effective drugs, which we will not go into here, are also available for the treatment of infectious diseases caused by fungi, protozoa, and helminths.

The backlash – anti-antibiotics

The golden age of antibiotics lasted from 1945 until the 1960s. Today the antibiotic weapons at our disposal have become blunt. Bacteria have "learned" to escape the effects of antibiotics and become resistant to them by undergoing mutation. The situation is compounded by the fact that the pipelines for new antimicrobials at most of the big pharmaceutical companies have run dry. One might ask what the big deal is about antibiotic resistance. After all, several drugs having distinct mechanisms of action are available to fight any bacterium. But often doctors have to contend with multiresistant bacterial strains that respond poorly, if at all, to any antibiotic therapy. You may be forgiven for thinking that synthetic antimicrobials are less susceptible to resistance development, since bacteria have never encountered them in nature. Yet some bacteria have already adopted effective counterstrategies against them.

Resistance development is in itself entirely normal. It is an inherent feature of bacterial evolution. Today for every antibiotic in existence, resistant microbes exist. Often there is a window of just a few years from the introduction of a new antibiotic until its effectiveness begins to wane. Even penicillin, originally a wonder drug, quickly lost its luster, when the first pus-forming bacteria (staphylococci) became resistant to it just a few years after its introduction. By 1960 penicillin was already powerless against nearly all staphylococcal strains. Doctors then reached for the next weapon. But this knee-jerk reaction of resorting to the next available antimicrobial is no longer working. In many hospitals patients face multiresistant staphylococcal strains, against which few weapons are available. In such cases vancomycin is often the antibiotic of last resort, but since 2000 vancomycin-resistant staphylococci have also begun to emerge.

Staphylococci are just one example of this trend. There are many others. There has also been an alarming increase in resistant *Streptococcus pneumoniae* (pneumococcal) strains. In the Czech Republic 60% of all pneumococcal strains are resistant to penicillin; in South Korea the corresponding figure is 70%. Antibiotics that are still effective tend to be newer drugs that are often far more expensive and are associated with more severe side effects than the older drugs. In many cases they lead to a hundred- or even thousand-fold cost explosion. For example, in Western Europe and the USA a tubercle bacillus that is susceptible to conventional drugs costs around 500 to 1000 euros to treat. Counting hospitalization costs and other factors, that figure can easily rise to more than 10,000 euros. In the case of tuberculosis too we have reached a crisis. Extensively drug-resistant tuberculosis (XDR-TB) strains that are resistant to conventional antibiotics have so far been reported in fifty countries.

The spread of antibiotic-resistant microbes is alarming. It is mainly due to the inflationary use of antibiotics. It is thought that up to 50% of all antibiotic prescriptions in industrialized countries are unnecessary – not only unnecessary, but dangerous, because the lavish use of antibiotics is the driving force behind the development of new resistance. If nothing is done to counter this trend, we will find ourselves returning to the preantibiotic era, which was before 1945.

Learning from resistant strains and winning

Bacteria develop resistance as part of their rapid evolution. They acquire this ability through mutations that confer a selective advantage over antibiotic-susceptible strains, enabling them to proliferate more rapidly. Alternatively, they may acquire resistance genes directly from other resistant bacteria, along with all the information they need to synthesize modified molecular channels, new enzymes, or foreign transporter molecules.

We live in a sea of bacteria, a surprising number of which inhabit our intestines (10^{12}–10^{14} bacteria). When we ingest resistant bacteria, they indulge in a veritable gene-swapping orgy with closely related strains that make up our intestinal flora. In so doing, they pass their resistance on to the intestinal flora. If it is an antibiotic-sensitive bacterium that causes diarrheal disease – for example a *Shigella* species that causes bacterial dysentery – it may acquire this dangerous ability from the intestinal flora. As early as 1959 several strains were found that were suddenly resistant to four antibiotic groups: tetracyclines, sulfonamides, streptomycin, and chloramphenicol, rendering those drugs ineffective. Today nearly all *Shigella* species in Southeast Asia are resistant to three or more antibiotics.

Every trick in the book

Bacteria employ three basic defense strategies against antibiotics: They can block out an antibiotic; they can immediately catapult it out again; or they can destroy it. In the first strategy, for example, a bacterium can modify the site at which the antibiotic acts. Microbes have accomplished this trick against almost every antibiotic we have. Alternatively, they can render their cell wall or cell membrane impenetrable to antibiotics. Many antibiotics are shuttled into the bacterium through molecular channels. If these are altered, the antibiotic will no longer be able to gain access. Strategy two – shuttling the antibiotic out of the cell again – relies on special transport mechanisms. The bacteria usher the antibiotic molecules back out, usually before they have a chance to do any harm. This is how a number of bacterial species that cause diarrhea deal with tetracyclines. (Incidentally, some cancer cells employ similar mechanisms to gain resistance to chemotherapy drugs.) Finally, the third mechanism involves, literally, cutting the enemy to pieces. Some bacteria accomplish this by producing enzymes that cleave the antibiotic molecule, rendering it inactive. One such enzyme is beta-lactamase, which many bacterial strains produce to destroy penicillin and related beta-lactam antibiotics.

6.2 When hospitals cause sickness: nosocomial infections and antibiotic resistance

They breed in the gaps between tiles on the walls of intensive care units or on dust-laden surgical lamps: hospital bacteria thrive anywhere that is supposed to be, but is not quite, sterile. For bacteria, the hospital environment is extremely hostile. Antibiotics – including valuable last-resort drugs – are everywhere in the air, and surfaces are constantly being sprayed and wiped with disinfectant solution. Unfortunately, the old adage holds true: what doesn't kill you makes you stronger – and more deadly. Hospital bacteria often prove to be resistance virtuosi. Most have learned to fend off assaults from outside and to prevail over the hostile environment. In some places in Germany up to 60% of all pathogens which patients contract in the hospital setting are resistant to at least some antibiotics and in many cases to several at the same time. In the USA, almost every pathogen encountered in hospitals is resistant to at least one antibiotic drug.

Hospital-acquired infections, known as nosocomial infections, are a growing problem in industrialized countries. Of every one hundred hospitalized patients, five to ten contract a bacterial infection they did not have when admitted. In Germany, half a million people are affected every year. In the USA, two million people acquire nosocomial infections every year, while in the UK, one in ten patients becomes infected with a nosocomial pathogen. Places posing a high infection risk are neonatal units, surgical wards, intensive care units, and, most critically, burn units.

It is a dangerous mix. One in two incidents in which hospital patients develop complications are due to nosocomial infections. All too often such infections are fatal: in Germany hospital-acquired infections claim 15,000 lives a year, while in the UK, 10,000 and in the USA more than 100,000 patients die of hospital-acquired infections. Around 70% of all deaths in US intensive care units are due to severe blood poisoning (sepsis) or septic shock. Bacteria invade the bloodstream and are transported with the blood to internal organs, where they can wreak mortal damage. In septic shock, the body's immune cells respond to the blood-borne bacteria by releasing excessive bursts of defense chemicals. This so-called cytokine storm is an example of phenomena in which the immune response can be fatal. The defense chemicals trigger massive inflammatory responses and damage the body's network of fine blood vessels, including those in organs. Sometimes they cause the blood in the entire body to clot and often prove fatal. (In the devastating drug test conducted by the company TeGenero in 2006 which resulted in six British trial subjects fighting for their lives, organ failure was very likely caused by a cytokine storm. All six men survived.)

Nosocomial infections are also catastrophic from an economic point of view. They cost industrialized countries up to 30 billion dollars a year: between five and ten billion dollars in the USA alone and around 1.5 billion euros a year in Great Britain.

Who, how, what?

Nosocomial microbes are usually bacteria that many people harbor on their skin or in their intestines, but without causing any problems. In hospitals, though, they invade weakened patients and can become a threat. Methicillin-resistant *Staphylococcus aureus* (MRSA) strains, vancomycin-resistant enterococci (VRE), and multiresistant strains of *Escherichia coli*, *Klebsiella pneumoniae*, and *Pseudomonas aeruginosa* are of especially grave concern. The list of disorders they cause include pneumonia, urinary tract infections, sepsis, local skin infections, and deep-tissue infections.

About one in three hospital-acquired infections is due to methicillin-resistant *Staphylococcus aureus*. The name does not imply that the bacterium is resistant to methicillin alone, it is resistant to *all* antibiotics *including* methicillin, until recently the only remaining weapon available against resistant strains. The first cases of MRSA were reported in 1968. Today, every year the note "MRSA" is written in the records of 40,000 to 50,000 patients in Germany. Only a quarter of patients survive MRSA blood poisoning. Yet Germany is still relatively well off in this respect, with less than a fifth of all *Staphylococcus aureus* cases exhibiting methicillin resistance. In Greece half of all staphylococcal strains are resistant. In England it is over 40%, in Italy somewhat less than 40%, and in France nearly 30%. In the USA in 2007 nearly 100,000 people were infected with MRSA, and 20,000 died as a result. The Scandinavian countries and the Netherlands have largely overcome the problem, reporting only 0.2% cases of methicillin-resistant staphylococci. In these countries MRSA patients are immediately sequestered to prevent the bacteria from spreading through the hospital. Thus, the problem can be successfully tackled through stringent hygiene measures.

The situation with so-called vancomycin-resistant enterococci

(VRE), the second most common agents of nosocomial infections, is also dramatic. These bacteria are part of the normal flora in the intestines in humans and many pets and domestic animals. However, in gravely ill patients they can cause life-threatening urinary-tract infections, sepsis, or encephalitis. The enterococci's vancomycin resistance was probably transmitted from animals to humans. Until recently an antibiotic resembling vancomycin was mixed with animal feed to promote growth. *E. coli* is also a normal occupant of the intestinal flora in humans and domestic animals and has therefore had ample opportunity to develop multiple resistance. Invasive *E. coli* strains are increasingly being found to be simultaneously resistant to as many as three antibiotic groups. Another agent of nosocomial infections that should be mentioned is *Pseudomonas aeruginosa*, which, although not a normal colonizer of humans, proliferates profusely in wash basins, incubators, vases, towels, washcloths, and dishcloths. In the USA *P. aeruginosa* is responsible for more than 10% of all hospital-acquired infections. Cystic fibrosis patients, in whom mucus accumulates continuously in the lungs, are especially susceptible, and many develop pneumonia as a result.

6.3 Five kilograms of penicillin, please: antibiotics in animal breeding

In many industrialized countries less than half the antibiotic used is administered to humans. The other half is used in animal breeding and in the meat-processing industry. The sheer number of domestic animals present in some regions is mind-boggling: the USA, for example, is home to more than five times more domestic animals than humans. In China there are probably ten ducks, geese, or chickens for every human, and in the Netherlands there are six times as many chickens as humans. And a whole new food industry is emerging: commercial fish culture is growing by 10% a year globally and is expected to exceed beef production by 2010.

Antibiotics are used worldwide in animal production for three reasons:

- Vets prescribe and administer them to sick animals. However, it is estimated that less than 15% of all antibiotic used is used for veterinary purposes.
- About a third of antibiotic use serves to prevent infectious diseases.
- More than half of the antibiotic used is fed to animals as performance enhancers.

What are performance enhancers? They are feed additives used to accelerate the growth of calves, bulls, or turkeys. Often they are antibiotics used in low doses, a practice that was also common in the EU until the end of 2005. Why and how they work is not entirely understood. The fact is, however, that animals fed in such a way are often ready for slaughter earlier than their penmates who are not fed antibiotics. It was probably due to performance enhancers that staphylococci developed resistance to vancomycin, the last effective secret weapon, in Europe. This development is often cited as an example of pen-bred resistance. Vancomycin exhibits cross-reactivity with another antibiotic known as avoparcin. What this means is that if a bacterium is resistant to avoparcin, it is also resistant to vancomycin. Avoparcin was widely used as an additive in animal feed. This practice led to avoparcin-resistant enterococci, which were then also insensitive to vancomycin. The step to vancomycin resistance in staphylococci was a small one and was accomplished through gene transfer. Since 2003 we have seen the emergence of staphylococci that are unaffected by vancomycin and conventional antibiotics. The high percentage of vancomycin-resistant enterococci in the European population – up to 10% – was the main reason underlying the 1997 EU-wide ban on avoparcin in animal breeding. In 1998 four other performance enhancers were declared illegal in the EU, and an EU ban on the use of antibiotics in feed has been in place since 2006. Certain antibiotics are permitted only on poultry and rabbit farms in order to prevent infestation by two parasites (coccidia and histomonads).

Denmark, which banned avoparcin as early as 1995, is regarded as a pioneer in this respect. Initially, the bacteria were found on 82% of all slaughtered turkeys, chickens, and ducks in the country. After the ban was imposed, the incidence of vancomycin-resistant enterococci (VRE) fell dramatically. Three years later it plummeted to just 12%. However, the effect was not as marked in the hog population. Thus, we see that animals can act as a reservoir of antibiotic-resistant microorganisms, which are then often transmitted to humans.

A pen is like a hospital

Transmission is not a one-way street. A modern densely crowded pen is, just like a hospital, an ideal breeding ground for resistant strains. Within it, hoards of bacteria live in close quarters, and a number of different antibiotics are used. New microbes are introduced via humans, rodents, birds, insects, new animals or pets, vehicles, feed deliveries, or water. The bacteria then multiply in the pen or in the effluent. Time and time again, deadly resistances develop. The bacteria are then carried outside the farm on vehicles, in animal carcasses, in meat, and in sewage as well as via humans, rodents, and insects. They also adhere to meat and eggs and are transported onto fields in slurry, and thus contaminate vegetables, rivers, and even drinking water, to mention just a few of the possible transmission pathways. Airborne transmission, especially in crowded pig pens, is also possible.

7 Self-defense: Immunization

A scientist who is also a human being cannot rest while
knowledge which might reduce suffering rests on the shelf.

Albert Bruce Sabin (1906–1993),
discoverer of the live polio vaccine

Well, the people, I would say. There is no patent. Could you
patent the sun?

Jonas Edward Salk (1914–1995), discoverer of the inactivated polio vaccine
in response to a question about who owns the vaccine. Neither Salk nor Sabin
patented their polio vaccine.

7.1 Introduction

Whenever I'm besieged by begging children and teenagers on
my trips to Africa or Asia, the most painful sight is that of the
lame and crippled. Polio isn't always to blame, but it usually is. It
reminds me of my schoolfriend Walter. He became ill with polio
in the '50s and suffered terribly, lying for weeks in an iron lung,
a man-sized metal tank that helped him breathe mechanically,
because Walter's breathing muscles could no longer function on
their own. Then his legs became paralyzed, cruelly excluding
him from our sports games. That was about 50 years ago. Today,
polio has disappeared from Europe, although it broke out again

in the Netherlands in 1992 in a community that had refused immunization for religious reasons.

The medical world can point to similar successes with other diseases: wherever vaccination programs have been carried out over an extensive area, the diseases for which they were developed have been largely eradicated. Apart from polio this also applies to measles, mumps, rubella, whooping cough, diphtheria and tetanus (Fig. 18). Besides these basic vaccinations, immunizations against *Haemophilus influenzae* type b (Hib vaccine), hepatitis B, chickenpox, meningococci, and pneumococci are also recommended in Germany – and recently the immunization against cervical cancer has been added to the list for girls. In the USA immunization against hepatitis A, hepatitis B, influenza, and rotavirus are also recommended. On the other hand, the BCG vaccination program for tuberculosis has been largely discontinued in Western Europe, since the vaccine only prevents childhood tuberculosis, and then only to a limited extent. Unfortunately, we do not as yet have a vaccine for the most common Group B meningococcal strains, though vaccines are still available against cholera, yellow fever, hepatitis A, rabies and typhoid fever, for certain high-risk groups and travelers to endemic areas.

Paradoxically, the success of immunization programs is also the reason for setbacks: once diseases have been conquered, they are out of mind. It then becomes difficult to justify the direct benefits of immunization. Many diseases have only ostensibly disappeared, as evidenced by recent outbreaks of measles in Germany, the UK and the USA, to name a few. In any case, the cost-benefit calculation will always be in favor of immunization.

For every individual it is valid to say that, taking the extremely small risk of side effects into account, immunizations guarantee protection against severe and often fatal diseases. Public health services also benefit economically. For every euro spent

Incidence of childhood diseases after the introduction of immunizations against them

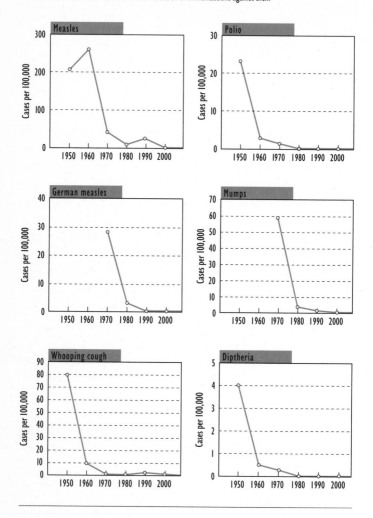

on immunizations against measles, mumps, or rubella, ten euros are saved that would otherwise go towards treating individuals with the disease. For every euro spent on immunizations against diphtheria, tetanus, or whooping cough, the system saves twenty euros. Of course, these are only rough estimates. The more prevalent a disease is in a country, the higher is the cost benefit factor. On the other hand, the financial gain in countries of the industrialized world, where many infectious diseases have become rarities – appears to be marginal at first glance. However, if such a disease were reintroduced, an unimmunized population would be at great risk, and the resulting costs would be astronomical. The benefits are much more evident in regions where infectious diseases amenable to immunization are still rampant. With regard to diseases such as polio and measles that have been largely eradicated, the cost-benefit relation has shifted, because we must continue to immunize worldwide even though the number of cases continues to fall. The benefit is only apparent when the disease is completely eradicated and the immunization program can be stopped. So far this has only been achieved with smallpox, and years of strenuous effort have paid off. The World Health Organization declared smallpox to be eradicated on May 8, 1980. Whether this success can be repeated is questionable. The next candidates on the list are polio and measles. The global polio immunization campaign has been running since 1988 and the measles campaign since 2001. Fewer than 2,000 children worldwide died from polio in 2006. The latest figures from the

Figure 18 Incidences of childhood diseases after the introduction of immunizations against them

The figure shows the drastic fall in case numbers following the introduction of general immunization against measles, polio, German measles, mumps, whooping cough, and diphtheria in various studies

World Health Organization show that fewer than 250,000 measles-related deaths occurred in 2006. The eradication of polio is proving to be more complicated than first thought. Polio flares up time and again, most recently in the border area between Niger and Nigeria. Thus, polio was not eradicated worldwide in 2005 as predicted by the World Health Organization. In 2006 the disease was reported in four countries: Afghanistan, Nigeria, India, and Pakistan. Efforts to eradicate measles, on the other hand, have had marked success, even surpassing the targets set.

From an economic point of view, the smallpox program was probably one of the most successful health investment programs the world has ever seen. Seldom has investment in health paid off so handsomely. The total cost of the program was some 300 million dollars. Before smallpox was eradicated, two billion dollars was spent annually worldwide on immunization, diagnosis, quarantine measures, and medical care – money that has been saved since 1980. In the USA alone, a total investment of 32 million dollars a month has been recouped. Other countries can point to comparable figures. A similar benefit has begun to emerge thanks to the program to eradicate polio. Once this target has been achieved, immunization against polio will save three billion dollars a year worldwide compared to the current situation. I believe a further argument is worth considering: the eradication of smallpox came just at the right time. It could not have been launched earlier, and it could not have taken longer than it did without dire consequences, because the HIV/AIDS epidemic followed close on its heels. This is because the smallpox vaccine is a live vaccine and is not without complications. Between ten and twenty cases of life-threatening complications and two deaths can be expected per one million immunizations, quite apart from the minor side effects. (A newly developed vaccine with such an adverse-effect rate would probably never be approved.) The adverse-effect rate in individuals with HIV/ AIDS would have been far worse. In the presence of a weakened immune system the vaccine virus could be virulent, causing severe complications. A smallpox vaccine could not have been given to HIV-infected individuals. Had the smallpox vaccination program lasted into the HIV/AIDS era, in all probability smallpox would never have been eradicated from the world's stage. And developing countries, particularly in Africa, would have had to bear an additional oppressive burden. We cannot afford delays in dealing with infectious diseases.

7.2 Vaccines for the masses

The smallpox eradication program provided valuable experience with mass vaccinations. It was no accident that the recommendation for an immunization program against childhood diseases was issued by the World Health Organization Smallpox Eradication Unit. In 1974 the World Health Organization and the United Nations Children's Fund called the Expanded Program on Immunization (EPI) into being. The aim was to provide every child on Earth with basic immunization during the first year of life. Whereas children in industrialized countries receive immunizations as a matter of course, those in many developing countries do not enjoy the benefits of national healthcare. The greatest number of deaths that could have been prevented by vaccination are reported in sub-Saharan African countries. Just under 60% of all whooping cough deaths, more than 40% of all deaths from tetanus, just under 60% of measles-related deaths, and the majority of all deaths from yellow fever occur here. Only in the case of hepatitis B do the majority of deaths (60%) occur elsewhere: in East Asia and the Pacific region.

As recently as the early 1970s, only one in twenty children in the world was immunized against polio, diphtheria, tuberculosis, whooping cough, measles, and tetanus. Since 1990, thanks in large part to the EPI, around 80% of children have received these basic vaccinations before their first birthday – some 100 million children a year. The result has been a fall in infant mortality from 17 million to 12 million. Last year alone the lives of up to 3.5 million children were saved. Countless others have been spared the harrowing consequences of those diseases. Nevertheless, 34 million children today still do not receive basic immunization.

A trivalent vaccine has been developed which confers protection against diphtheria, tetanus, and whooping cough. Known as the DTP vaccine, it has to be administered three times at four-week intervals. The BCG vaccine against tuberculosis is given right after birth. Although this is not an ideal solution, the practice prevents childhood tuberculosis to some extent. The oral live polio vaccine is used in developing countries. It is cheaper and safer than the inactivated vaccine given by injection. Moreover, the live vaccine inhibits replication of the virus in the gut, whereas the inactivated vaccine only prevents the disease. Thus the live vaccine simultaneously stops the spread of the polio virus. Children should be immunized four times ideally, but at least three times. This can be coupled with the BCG vaccination and the three DPT vaccinations. Children are immunized against measles at the end of their first year of life. These basic immunizations are inexpensive, costing about twenty dollars per child. It would therefore cost around one billion dollars to immunize every child in the developing countries.

If possible and where necessary, other vaccines, in addition to the basic ones, are available at considerably greater cost. Three injections are needed for protection against hepatitis B. Hepatitis B immunization has become more common since the price for vaccination through the public healthcare system fell recently. Immunization against yellow fever is available mainly in African and Latin American countries where the disease is endemic. Vaccinations against *Haemophilus influenzae* type b is becoming increasingly common, as are vaccinations against pneumococcal and meningococcal infections, though the costs are still extremely high. Vaccines against the human papillomavirus, the causative agent of cervical cancer, are much too expensive for widespread use in developing countries.

The Global Alliance for Vaccines and Immunization (GAVI) was founded in 1999. With its financial support, between 2000 and 2005 some 13 million children were vaccinated against DTP, 90 million children against hepatitis B, 14 million against *Haemophilus influenzae* type b, and 14 million against yellow fever. These vaccinations were administered safely thanks to 1.2 billion disposable syringes. GAVI is supported by, among others, United Nations Children's Fund, the World Health Organization, the World Bank, the Bill & Melinda Gates Foundation, some industrialized and developing countries, and several vaccine manufacturers. Some of the eleven supporting countries in Europe are Denmark, France, Ireland, Luxemburg, the Netherlands, Norway, Sweden, and Great Britain. Funding for GAVI is ensured until 2015. By then around 10 million lives will have been saved with the help of 500 million vaccinations.

7.3 Immunization risks: myths and truths

No vaccination is totally risk-free. The probability of serious complications is on the order of one per 100,000 to 1,000,000 vaccinations. The severity and frequency of complications varies depending on the vaccine. In Germany, suspected (note:

suspected not definite) complications occur in about three per 100,000 vaccinations. Thus, there were just under 1400 suspected cases of complications in 2005, of which 900 were severe. Thirty-four people suffered permanent disability, and twenty-three died. However, more thorough investigations showed that a coincidental rather than a causal correlation was more likely in nearly all the suspected cases. Among the 2630 reported suspected cases from 2004/2005, the investigation was unable to exclude an actual link with a vaccination in just two severe cases in adults. One case involved Guillain-Barré syndrome in a man with prostate cancer who had been vaccinated against influenza. In the second case, a forty-four-year-old man died of meningitis after receiving hepatitis and polio immunizations. In the two decades between 1988 and 2008, more than 12,500 claims were filed with the US National Vaccine Injury Compensation Program, of which around 2200 were deemed worthy of compensation.

Questions about vaccinations by concerned parents must be taken seriously and discussed in detail. At the same time, however, complications must be viewed in relation to the severity of the disease that the vaccine is meant to prevent. Numerous myths about supposed correlations abound, for instance that the hepatitis B vaccine triggers multiple sclerosis or that the mumps-measles-rubella vaccine (MMR) favors the development of autism in children. The list goes on: the *Haemophilus influenzae* type b and hepatitis B vaccines are said to cause type 1 diabetes; the rubella vaccine, arthritis; the influenza and meningococcal vaccines, Guillain-Barré syndrome; and the DPTH vaccine, SIDS (Sudden Infant Death Syndrome). Closer research has not found any correlation in any of these cases.

The greatest furor in this respect was caused by a study report published in the respected medical journal *The Lancet* in 1998. In the report, the authors claimed a link between the trivalent MMR vaccine on the one hand and bowel disease, developmental disorders, and autism on the other. After numerous secondary studies showed the claim to be untenable, all the authors retracted the conclusions except the lead author. He stood by his findings and was later accused of a serious conflict of interest, since he had received generous financial support from groups that wanted to take legal action against vaccine manufacturers after their children had developed autism. That doctor is currently on trial for fitness to practise by the British General Medical Council for serious professional misconduct.

Sometimes criticism of vaccines has been so wide off the target that constituents were blamed for side effects which the vaccine in question did not even contain. Thimerosal is a mercury compound which is added to vaccines as a preservative. It is used, for instance, in diphtheria, whooping cough, tetanus, and hepatitis B vaccines. It has never been added to live vaccines, such as those that protect against mumps, measles, and rubella. Thimerosal was purported to cause neurological developmental damage. The accusation is simply absurd when applied to live vaccines. Moreover, the amount of thimerosal in inactivated vaccines is too minute to cause any demonstrable damage. Nevertheless, basic immunization with thimerosal-free vaccines is now guaranteed throughout Europe. Since 2001, routine vaccinations in the USA are thimerosal-free. Thimerosal also came under suspicion following the introduction of a new, intranasal influenza vaccine, when frequent cases of facial paralysis were reported. The vaccine was immediately withdrawn. The cause turned out to be a new adjuvant.

7.4 Me and the rest of the world

Standard vaccinations are recommended for all the inhabitants of a country, even if the disease no longer occurs there. The purpose of vaccinations is not only to protect the individual but also the community as a whole. Only if the immunization rate is over 90% can we assume that any pathogen brought into a country (e.g. by tourists or immigrants) will not do much harm. One could take the flippant view that the 10% who have not been vaccinated will be fine as long as they do not travel to high-risk areas. Of course, in exceptional cases, even with an overall vaccination rate of 99%, an infected person could pass the infection on to his close family, i.e. his wife and children, if they have not been vaccinated.

All US states require children to provide proof of immunization before entering daycare and/or school, and must carefully adhere to a prescribed schedule of vaccinations. In contrast to the USA, vaccinations are recommended, but not compulsory in the UK or Germany, which is why the onus should be on the individual, particularly the parents of infants, who must make a decision after weighing all the available information. What is needed is the kind of immunity, known as herd immunity, that occurs when the immunization of a large percentage of the population (or herd) provides protection to individuals who cannot be vaccinated, e.g. immune-deficient patients and newborn infants. What is needed is precisely this kind of solidarity – not just in Germany but worldwide. This is why I am ardently in favor of vaccinations. They are the reason why many diseases seldom or no longer occur in the industrialized world. There is in fact a minimal risk of complications from vaccines, but it is a 100,000 to one million times smaller than that of the serious diseases they protect against. You are more likely to be hit by lightning. And let us not forget that every year vaccinations save more than five million lives. The day the diseases in question are gone for good, as in the case of smallpox, is the day we can consider suspending immunization programs.

8 Poverty and Infectious Diseases from a Global Point of View

Help those who cannot help themselves!

From the Congo

Alms destroy the soul of the donor and the receiver and also fail in their purpose, for they worsen poverty.

Fjodor Dostoevsky

8.1 Introduction

Epidemics are transnational: they respect no national territorial limits, and no country is able to stop them at its borders. The recognition that poverty, social inequality, and disease form a vicious circle, has motivated international organizations of various structures to provide considerable funds to combat them: more than 40 billion dollars since the turn of the century. A large part of this flows into programs set up to fight transmissible diseases, especially vaccination programs to prevent childhood diseases, and into programs aimed at controlling today's major HIV/AIDS, tuberculosis, and malaria epidemics. The focus of these activities has now shifted to sub-Saharan Africa, which is most acutely affected by these problems. No one can doubt that poverty contributes to the spread of infectious diseases. Dirty water and poor sanitation cause widespread diarrheal diseases. In many areas there simply are not enough mosquito nets to halt the inexorable march of malaria. Drugs and immunization programs are lacking to control major epidemics. Diseases then further weaken the fragile economies of impoverished countries, further exacerbating the plight of the poor.

8.2 Money, health, education

Definitions of poverty are legion. The most concise ones define poverty in terms of economics. In Europe the poor are deemed to be those who earn less than 60% of the average income in the country concerned. In the case of developing countries, the world's poorest are usually defined as those who have to make do on less than one dollar a day. One billion people fall into this category. The poor are those who have to manage on less than two dollars a day. (By comparison: In the EU every cow is subsidized to the tune of two euros a day.) Based on this yardstick, about half the world's population, i.e. around three billion people, are poor. Nine in ten people in the world have a disposable income of less than 5000 dollars a year.

Aid for Africa from industrialized countries has grown to 625 billion dollars since the Second World War – an enormous sum. However, it must be viewed in the context of other billion-dollar expenditures: every year wealthy countries pour half this sum into agricultural subsidization programs. At the same time, money spent in this way contributes to growing poverty in developing countries. When Europe or the USA send heavily subsidized food produced as a generous gift to developing countries, one effect is to undermine agriculture in the recipient countries. We not only need to organize development aid better but must also rethink the subsidy policies of industrialized nations. Please do not misunderstand me: sending food aid to disaster areas is the right thing to do. What is unfair, though, is the lopsided competition between farmers in developing countries and subsidized agriculture in industrialized countries. Indeed, there are signs that a fundamental rethink is in progress: Canada supports the purchase of food from local farmers or, in crisis situations, the distribution of money directly to the population, enabling people to buy food locally. The European Union is also taking a new tack. Only 10% of food aid can come from the EU's own stockpiles. Now the USA, which is by far the biggest food donor, must follow suit.

Economic criteria are useful for defining poverty, but they fail to do justice to the full complexity of the problem. They do not consider, for example, factors such as nutrition, health, and education. Nearly one billion people worldwide suffer from hunger and malnutrition, which claim 30 million lives a year. Every year 20 million children are born underweight, which often has life-long consequences. Nearly a third of all people who do not live in cities have no access to clean drinking water. In sub-Saharan Africa fewer than half of the rural population receives clean water, whereas in the cities over 80% do.

It may sound trite, but the truth remains: we live in a divided world in which 20% of humanity enjoy plenitude and generate 80% of the global gross domestic product, while the other 80% live in abject poverty. Yet food and health are basic universal human rights guaranteed, not least of all, by the United Nations Charter.

8.3 The *Who's Who* of organizations

Originally, development aid was a matter for taxpayers. They were the source of funds which governmental and intergovernmental organizations managed and disbursed to foreign governments. Today, however, the most successful organizations at work are multinational hybrids made up of governmental, intergovernmental, nongovernmental, civil, and private groups which are able to act very flexibly.

Governmental organizations (GOs) and intergovernmental organizations (IGOs) involve institutions or persons of the state, including governments, while nongovernmental organizations (NGOs), civil organizations, and private organizations are independent of governments. NGOs may also include organizations with financial interests, for example, associations of the pharmaceutical manufacturers. Private organizations are either private foundations, such as the Bill and Melinda Gates Foundation, or commercial entities such as pharmaceutical companies. A civil organization is an NGO serving the public interest. Finally, private-public partnerships (PPPs) are partnerships of civil and private organizations serving the public interest.

8.4 Economic strategists

Economists see a key motivation for development aid in the effect it has on the global market economy. Disease is viewed mainly as an economic factor, the fight against it as a contribution to economic strength. This view now permeates our understanding of global health as an economic factor and has helped mobilize resources to remedy the current situation. However, the market-biased view has also created problems, in particular in matters relating to patent law and the distribution of life-saving drugs at affordable prices in developing countries. A purely economic but nevertheless intriguing view is described in the box entitled "Thinking like a do-gooder."

World Trade Organization (WTO)

The aim of the WTO is to improve international trade and economic relations. The WTO emphasizes the liberalization and deregulation of trade policies with maximum levels of privatization. Currently the WTO has 150 members of varying economic clout: developing countries, threshold countries, and industrialized nations. The realization that the gulf between rich and poor countries is widening has increasingly prompted the WTO to raise inequality issues, even if the supposed benefit to developing countries is not always clear.

International Monetary Fund (IMF)

The IMF represents 185 members. Their voting rights, however, are based not on demographic principles but rather on capital quotas. The countries with the greatest capital, i.e. the USA, Japan, Germany, France, and Great Britain, have a voting quota of nearly 40%. The EU members together have a quota of over 30%. The IMF aims to promote international trade by stabilizing exchange rates and regulating monetary policies. Strict regulations on lending have led to a deterioration of the healthcare system in debtor countries. After the breakdown of the Soviet Union, Eastern European countries and former member states of the Soviet Union participated in IMF lending programs to better their economic situation. The financial reduction for tuberculosis programs requested by IMF markedly aggravated the tuberculosis burden in these countries. The current president of Germany, Horst Köhler, was the managing director of the IMF from 2002 to 2004.

The World Bank

Even more so than the IMF, the World Bank pursues the stated goal of reducing poverty by promoting economic development. Its most important goals are to improve education, healthcare, agriculture, and environmental protection and to combat corruption. With its 1993 report "Investing in Health," the World Bank was instrumental in promoting the perception of health issues as a global problem. In May 2007 former President of the World Bank Paul Wolfowitz was forced to resign due to personal affairs.

This enabled the US President to propose the man to succeed him. The new man is Robert Zoellick, a long-time crony of President Bush. It is interesting to note that the highest office in the World Bank, which demands competition, performance, and transparency from developing countries, was filled while circumventing precisely these mechanisms.

An important function of the World Bank is to grant loans to countries in precarious economic circumstances. Every year the World Bank lends 20 billion dollars – often with strict conditions attached and in many cases with harmful effects. For example, Costa Rica received money on loan with the proviso that it institute cost-saving measures, whereupon it made cuts to its healthcare programs. As a result, infectious diseases and child mortality soared. Responding to constraints imposed by the World Bank, China starting charging its tuberculosis patients for treatment, 1.5 million tuberculosis patients were forced to forgo treatment, and an additional 10 million people contracted the disease.

Instead of imposing sweeping cost-cutting stipulations, the World Bank and the IMF could make debt-relief measures contingent upon undertakings to improve healthcare.

Thinking like a do-gooder

An ecological *enfant terrible* or one of the world's most influential thinkers? Both epithets have been applied to Bjørn Lomborg, a young Danish professor of statistics. In 2001 he was catapulted to public attention with the publication of his book *Apocalypse No!*. Wielding a sharpened pencil, he calculated that the catastrophes so often invoked by doomsayers are not actually so dire after all, that Armageddon is not imminent, and that things in fact are steadily improving for humanity. Environmental groups groaned. In 2004 Lomborg invited eight of the world's leading economists, four Nobel laureates, and 30 experts in various fields to a roundtable discussion of a single question: How could 50 billion US dollars be used most effectively to improve the world? The answer is simple: by setting priorities. But where exactly? The scientists did cost-benefit calculations and issued a list of recommendations. The Copenhagen Consensus confirms that a great deal can be accomplished in the fight against major infectious disease with relatively little money. The top four recommendations in terms of their cost-benefit ratings were:

1. The fight against HIV/AIDS. An investment of 27 billion dollars could prevent nearly 30 million new infections by 2010. The experts calculated a cost-benefit factor of 40, meaning that for every dollar invested, the world would get 40 dollars in return.
2. The fight against hunger and malnutrition. Dietary deficiencies of iron, zinc, iodine, and vitamin A increase susceptibility to infectious diseases, especially in young children. The Copenhagen group would invest 12 billion dollars in this area.

3. Trade liberalization costs little but could yield large benefits, the experts concluded. This is a controversial issue.
4. Malaria control: 13 billion dollars for impregnated bed nets.

The cost-benefit ratio was far less favorable for global warming projects. Once again, environmental organizations foamed at the mouth. The approach, they claimed, translates human suffering into a financial quantity and grossly simplifies complex problems. For many, the Copenhagen Consensus was heresy. In reality it was nothing less than an attempt to find clear solutions to real problems.

A second conference followed in 2006, to which representatives of the United Nations and other diplomats were invited. A representative assortment of participants from developing, threshold, and industrialized nationals drew their conclusions. The number one priority, they thought, was to improve basic healthcare. Priority 2 went to the provision of basic sanitation and clean water. Control of HIV/AIDS and malaria came sixth and seventh, respectively. (Proposals to curb climate change again trailed behind because they would be too expensive.) The next round took place in 2008. For five days the question mooted by the panel was: "If you had an extra 75 billion dollars to do good in the world where would you spend it?" Micronutrients and food additives to combat malnutrition in the 140 million undernourished children reached position one and three. The second position considered implementation of international trade regulations to the benefit of the developing world. The fourth rank was given to improve childhood immunization programs.

8.5 TRIPS: patent rights versus treatment rights

A key economic issue in the fight against infectious diseases is the right to patent vital drugs and vaccines in developing countries. Any nation that seeks entry to the WTO must recognize its 1995 Agreement on Trade-Related Aspects of Intellectual Property Rights, or TRIPS for short. Article 27 of the TRIPS Agreement underscores member states' right to intellectual property and the protection of that right by means of patents. At the same time, Article 8 recognizes that member states are entitled to take measures to maintain the health of their population. This has led to a debate as to whether poor countries are allowed to manufacture and market drugs cheaply by circumventing patent and licensing rights. To clarify this point, a conference was held in Doha, Qatar in 2001. Although the first round ended without any clear resolutions, a declaration was framed, according to which every member country has the right to protect the health of its population, which includes providing access to drugs at affordable prices. Thus, it is possible to import generics from abroad, even if the drugs are patented. Although the Doha round did not amend the TRIPS Agreement, and patent protection continues to apply to all medications, thanks to the Doha Declaration it is now possible for countries to suspend patent rights and implement compulsory licensing.

When South Africa passed a law in 1997, i.e. before the Doha Declaration, the aim of which was to ensure access to affordable HIV/AIDS drugs (via parallel imports, generics, price controls), thirty-nine pharmaceutical companies brought the case before the country's Supreme Court of Appeals. However, before the court reached a decision, the plaintiffs realized that they had miscalculated. The international outcry was so loud that the pharmaceutical companies withdrew their appeal in 2001 and negotiated an agreement with the South African government. According to the terms of the settlement, South Africa is allowed to import and market generic AIDS drugs, but it acknowledges the general validity of TRIPS.

Invoking the Doha Declaration, Brazil's president Lula da Silva announced in May 2007 the compulsory licensing of an AIDS drug manufactured by Merck & Co. Brazil has long offered AIDS drugs free of charge to all individuals with HIV/AIDS with considerable success. Although Merck declared that it was prepared to cut the price of the AIDS drug, it was still significantly more expensive than the generic product. When Thailand imposed compulsory licensing of an AIDS drug manufactured by Abbott Laboratories in January 2007, Abbott withdrew its license applications for seven drugs in Thailand, and the US government issued a warning to Thailand about its disregard of intellectual property rights. The Swiss pharmaceutical company Novartis is currently locked in a dispute with the Indian government about the production of affordable generic AIDS drugs. These examples illustrate the dilemma the TRIPS Agreement entails – even after Doha: On the one hand, South Africa, Thailand, and Brazil are entitled to order the compulsory licensing of vital drugs. On the other hand, the TRIPS Agreement demands the protection of patents wherever possible. The pharmaceutical multinationals cite the high costs for developing new drugs. They are also supported by governments, whose coffers are swelled by the huge amounts paid in taxes by pharmaceutical companies, most notably by the US administration.

In the debate about who is right, it should also be mentioned that some pharmaceutical manufacturers have set laudable precedents. Gilead, for example, is providing an HIV/AIDS drug to the 100 poorest countries at an affordable price with a 5% profit margin. Bristol-Myers Squibb ceded its production and marketing rights to generic manufacturers in sub-Saharan Africa and India without levying any licensing fees. The upshot is that the TRIPS Agreement needs to be adapted to reality on the ground. Existing patent laws must be relaxed in the case of urgently needed drugs.

It is all reminiscent of the wake of the anthrax attacks in the USA in 2001. When the German pharma company Bayer AG did not immediately supply large quantities of Ciprobay, the only drug approved in the USA for the treatment of anthrax, the US government was not at all squeamish about making the company see reason: it threatened to suspend the drug's patent protection and manufacture it in the USA. The threat worked. Ciprobay was soon available at half the original price. The government's justification, incidentally, was that the drug was vital for the protection of its citizens.

8.6 Flagships of healthcare improvement

United Nations Organization (UN/UNO)

The UN is an international community comprising 192 countries. Its goal is to promote world peace, international law, and human rights. At the start of this century the UN announced the Millennium Development Goals (or Millennium Goals for short), which were called into being by the then general secretary, Kofi Annan. The goals are outlined in Section 8.7.

World Health Organization (WHO)

This specialized agency of the United Nations, which currently has 193 member states, has an annual budget of 1.8 billion dollars. That is less than New York City's annual budget for street cleaning. The objective of the WHO is to ensure the attainment by all peoples of the highest possible level of health that will enable them to pursue socially and economically productive lives.

The WHO is particularly committed to combating infectious diseases and has implemented a number of vaccination programs, many of which are funded jointly by private- and public-sector sponsors.

United Nations Children's Fund (UNICEF)

UNICEF focuses on children in need, originally providing help to children in countries devastated by World War II. Today it mainly benefits children in developing countries. In 1965 UNICEF was awarded the Nobel Peace Prize for its work. UNICEF organizes many healthcare projects, especially vaccination programs, in cooperation with the WHO.

8.7 Ambitious goals

On September 8, 2000 the UN member states adopted the Millennium Declaration, setting out goals to eradicate poverty and inequality, improve health, ensure a sustainably clean environment, and promote a fairer and more peaceful world. Many of the specific goals relate to poverty-associated infectious diseases.

The Millennium Goals seek to reconcile social and economic conditions, basic medical care, and major causes of hunger, inequality, and disease. It may prove to be an attempt to square the circle. The member states not only all signed the declaration but also promised to achieve the goals by 2015. It is arguable whether the promises were bold or reckless. In any case, it is now obvious that the Millennium Goals cannot be attained unless every effort is made to make up for lost time. Based on the status quo in 1990, the most important goals to be achieved by 2015 are:

Goal 1: Eradicate extreme poverty and hunger. Concretely, the proportion of the world's people whose income is less than one dollar a day is to be halved. The number of people who suffer from hunger, which now stands at 600 million, is also to be halved.

Goal 2: Achieve universal primary education for boys and girls alike by 2015.

Goal 3: Promote gender equality and strengthen the rights and role of women, especially at the educational level.

Goal 4: Reduce under-five child mortality by two-thirds of its current rate and markedly increase the immunization of young children against measles.

Goal 5: Reduce maternal mortality by three-quarters, not least of all through improved medical care of women during pregnancy and childbirth.

Goal 6: To halt and if possible start to reverse, the spread of AIDS, the scourge of malaria and tuberculosis, and other major diseases that afflict humanity.

Goal 7: Ensure environmental sustainability. Concrete goals in this respect include halving the proportion of the world's people who are unable to reach or afford safe drinking water and basic sanitation and to improve the living conditions of at least 100 million slum dwellers by 2020.

Goal 8: Develop a global partnership between individual states. Important goals in this respect are to improve trading and financial systems, enhance debt-relief programs, improve working conditions, and ensure free access to new technologies. For combating infectious diseases it is especially important that essential drugs are available in developing countries at affordable prices.

8.8 Theoretically no problem

To aim beyond mere declarations, the UN General Secretary at the time, Kofi Annan, set up a working group, the Commission for Macroeconomics and Health, whose brief was to show how the goals could be attained by 2015. It was also charged with determining whether the required investments by the rich to solve the health problems of the poor would have direct economic benefits, i.e. whether the required capital would be well invested. The head of the Commission was Jeffrey D. Sachs, Director of the Earth Institute of Columbia University in New York and professor for sustainable development and healthcare policies. Before Sachs became Annan's special consultant for the Millennium Goals, he had already advised numerous administrations and nongovernmental organizations. However, as an advisor to President Boris Yeltsin, his attempt to promote a free market economy in Russia failed miserably.

On December 20, 2001 Sachs submitted his report to the then director of the WHO, Gro Harlem Brundtland. The problem child – the report stated – was sub-Saharan Africa. Improved health is a prerequisite for economic development in this region. The report pointed out that many of the problems could be tackled just by improving the public health system. Sachs proposed that poor countries should contribute according to their financial means; the bulk, however, must come from the industrialized countries. A total investment of around 30 billion dollars, about 0.1% of the combined gross domestic product of the industrialized countries, would be needed. These calculations do not even include the special problems caused by HIV/AIDS, which would increase the amount needed by two billion dollars. That is the amount of money required to save eight million lives and 330 million DALYs – 2500 dollars per human life. The experts also calculated the expected financial returns: up to 360 billion dollars could be generated in this way.

8.9 Update I: déjà vu?

The Millennium Development Goals are not the first attempt by the international community to develop solutions to the world's health problems. As early as 1978, 134 delegates at a conference in Alma Ata (now Almaty), Kazakhstan, appealed to the WHO and UNICEF to secure the health and well-being of the world's population by 2000. Back then, the eradication of smallpox was almost a *fait accompli*, a proof of the potential of medical interventions. War, hunger, inequality, and poverty were seen as important causes of disease. The Alma Ata Declaration emphasized the right of individuals to basic healthcare and held states responsible for maintaining the health of their citizens. By 2000 everyone was supposed to be healthy and leading a socially and economically productive life. The goals were not even remotely realized.

In 1986 the International Conference on Health Promotion in Ottawa, Canada took the same line. Most of the participants still did not realize the scale of the HIV/AIDS catastrophe looming ahead. Many hoped that the aims formulated in the Alma Ata Declaration could be realized. Again, the focus was on the social conditions conducive to a healthy life: peace, security, education, a reliable food supply, a steady income, a stable ecosystem, rational use of resources, and equality and justice. Perhaps the discussion lost its way at this point. It had been forgotten that pathogens, lifestyle, and an individual's genetic constitution are also factors in the development of diseases. Just as combating diseases is not in itself sufficient, merely changing the social conditions is also an inadequate response.

In 1993 the World Bank published a report entitled "Investing in Health" which factored health problems into the World Bank's economic plans. This was a seminal document, because it firmly established the interaction between poverty and disease in the minds of decision-makers.

In 1997, eleven years after Ottawa, the Fourth International Conference on Health Promotion was held in Jakarta. Although the tone of the previous conferences was maintained, disease-specific factors figured more prominently in the Jakarta discussion. For the first time experts called for a global health alliance to devise and finance global health strategies. Even the industrialized countries appeared to realize that mobility in a globalized world applies not only to goods and labor but also to pathogens. HIV/AIDS had developed into a global threat, antibiotic-resistant bacteria were spreading around the globe, and tuberculosis and malaria were resisting all attempts to stamp them out.

8.10 Update II: mid-point review

The year 2005 marked the one-third point for realizing the Millennium Goals. It was time to take stock. Jeffrey D. Sachs remained stoic. "In 2005," he said, "the international community must try vigorously to meet its own commitments, taking quick practical steps before the goals become impossible to achieve. […] If we don't act now, the world will live without millennium goals, and it will be a very long way to the next millennium summit in the year 3000."

Two and a half years later, in July 2007, the halfway point for the Millennium Goals arrived. The picture had hardly changed. On July 2 the UN presented the "collective mixed record", as it put it. The diplomatic but clear conclusion drawn by General

Secretary Ban Ki-moon: "The results show that success is still possible in most parts of the world, but they also point out how much remains to be done." [...] "The world wants no new promises," he wrote in his foreword. Most of the highly developed countries failed to meet their financial commitments. At their summit in Gleneagles in 2005 the leading industrialized nations had pledged to double aid to Africa by 2010, it is stated in the summary. "But between 2005 and 2006 official aid declined by 5.1 percent." Only five donor countries had achieved the UN goal and now allocate 0.7% of their gross national income to achieving the Millennium Goals. With regard to the individual goals the UN made the following midpoint assessment:

Goal 1: The UN sees marked progress on the way towards halving extreme poverty. The number of people living on less than one dollar a day fell from 1.2 billion in 1990 to 980 million in 2004. However, the goal appears not to have been met in sub-Saharan Africa.

Goal 2: Good grades for the education offensive: In 2005, 88% of all children were enrolled in primary education. But girls and children in rural areas are still least likely to attend school. Sub-Saharan Africa remains a problem area.

Goal 3: Slow progress towards equality. The proportion of women in employment increased slightly, and women's political participation is also growing in many regions, albeit slowly.

Goal 4: Child mortality has fallen. In 2005, 10.1 million children died before their fifth birthday. Preventable diseases, especially AIDS and malaria, remain the chief causes of child mortality.

Goal 5: The goal of reducing maternal mortality by three-quarters appears impossible to achieve as things stand. Half a million women continue to die each year during pregnancy or childbirth.

Goal 6: The incidence of tuberculosis has begun to fall, but the absolute number of new cases is still rising due to population growth. The disease looks set to decline. The situation with HIV is sobering. Deaths from AIDS rose to 2.9 million in 2006. HIV prevalence has stabilized in developing countries but continues to rise in sub-Saharan Africa. Preventive measures are failing to keep pace with the spread of the disease. Malaria control measures are beginning to pay off, but few countries have achieved the targeted 60% coverage of insecticide-treated bed nets.

Goal 7: Half the world's population still has no access to basic sanitation. The rapid development of cities is making it an even more daunting challenge to improve the living conditions of slum dwellers. In addition, the effects of climate change are being felt, especially in poor regions. Climate change is a serious obstacle to achieving the Millennium Goals.

Goal 8: Debt relief of some countries has certainly helped improve global partnership. However, the majority of donor countries have failed to allocate the promised amounts. The World Bank, contrary to its promise, had not succeeded in clarifying pressing trade issues by 2007, which would benefit developing countries. In some cases import duties and trade obstacles have been dismantled, but not always with wholesale benefits for poor countries. In many countries gains from economic growth have not been equitably distributed. The number of unemployed young people rose to 86 million in 2006.

Other groups judge the progress made towards the Millennium Goals much more critically than the UN report. Take, for example, the figures on disease control published in *Le Monde Diplomatique* (Figs. 19 to 22).

8.11 Policymakers

Organization for Economic Cooperation and Development (OECD)

The OECD is an organization that aims to promote economic development. Nearly all thirty member states are industrialized countries, and this is reflected in the OECD's outlook: the organization mainly represents the interests of its industrialized members. The OECD also wants to boost economic growth in developing countries, which includes tackling disease-related problems. At present the OECD provides over 100 billion dollars in support to developing countries. The Millennium Development Goals were formulated on the basis of the OECD's report "International Development Goals."

G8

In a sense an elite faction within the OECD, the Group of Eight (G8) comprises the seven leading industrialized countries and Russia. Germany, France, Great Britain, Italy, Japan, Canada, and the USA, with less than 15% of the world's population, together account for two-thirds of global trade and gross world income. The G8 serves to coordinate and plan financial and currency issues. However, in recent years the member countries have also increasingly taken a stance on global issues, including health problems and infectious diseases. At its 2005 summit at Gleneagles, Scotland, the G8 recognized that Africa can realize its enormous potential only if widespread infectious diseases are effectively controlled. At the 2006 summit in St. Petersburg infectious diseases were again broadly discussed. The 2007 G8 summit at the Baltic spa town of Heiligendamm, which was hosted by German Chancellor Angela Merkel, emphasized improvements in healthcare systems and the control of HIV/AIDS, tuberculosis, and malaria in connection with growth and responsibility in Africa. At the end of the summit, the German Chancellor pledged 60 billion dollars from the G8 members for fighting AIDS, tuberculosis, and malaria to be paid, however, within a vaguely defined period of time.

Unfortunately, the G8's declarations of intent are not binding or legally enforceable. Usually they peter out at the level of announcements, and the summits are soon followed by disillusionment. At Gleneagles, Africa was promised around 50 billion dollars a year by 2010. Only 10% of that has so far been received. And only Chad and Cameroon have actually been relieved of debts, as emphasized by the chair of the Africa Progress Panel, Kofi Annan, when he met German Chancellor Merkel in the runup to the G8 summit in April 2007 and as underscored by the new UN General Secretary Ban Ki-moon in the 2007 update on the Millennium Goals. The 2008 G8 summit in Hokkaido Toyako, Japan reiterated all of these commitments with little sign of true action or breakthrough.

Up to now, donor countries have been able to present good figures relatively easily. Relief on debt has been reported as expenses, but the relief had already been written down and does not affect annual financial budgets. If we subtract debt relief, financial support for Africa from the OECD declined in 2006. If we seriously want to combat poverty actively and thus counter the threat of infectious diseases, the target is 0.7% of gross national income (GNI) in development aid. Germany is currently giving around 0.36%, or just 0.3% discounting debt relief. Yet

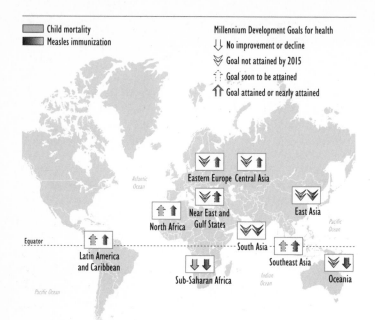

Figure 19 Interim progress made towards the Millennium Development
Goals for health in 2005: child mortality and measles
immunization
From Atlas of Globalization, *Le Monde Diplomatique* (2006)

at the G8 summit in Heiligendamm, Germany indicated that
it would like to contribute 0.5% of its GNI in the next three
years and 0.7% by 2015 to development aid. The firm pledge to
increase the budget of the Ministry for Economic Cooperation
by 750,000 euros at least raises an expectation that development
aid from Germany will reach 0.5% of the country's GNI.

In 2006 the USA spent only 0.17%, Australia 0.30%, and the

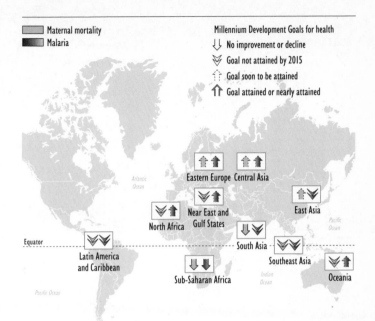

Figure 20 Interim progress made towards the Millennium Development Goals for health in 2005: maternal mortality and malaria
From Atlas of Globalization, *Le Monde Diplomatique* (2006)

UK 0.52% of GNI on development aid for developing countries. Model examples include Sweden, which contributed over 1.0% of its GNI to development aid, followed by Luxembourg and Norway (0.89%), the Netherlands (0.81%), and Denmark (0.80%).

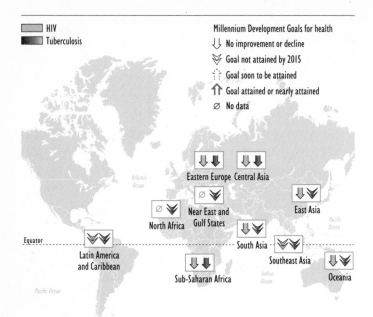

Figure 21 Interim progress made towards the Millennium Development
Goals for health in 2005: HIV/AIDS and tuberculosis
From Atlas of Globalization, *Le Monde Diplomatique* (2006)

8.12 Finger in the wound

DATA

DATA stands for Debt, AIDS, Trade, Africa. The organization
is publicized by, among others, the musicians Bob Geldof, Bono,
and Herbert Grönemeyer. According to DATA, the G7 states
(Russia refused to pledge any funds) have increased their development aid by 2.3 billion dollars, but there remains a shortfall

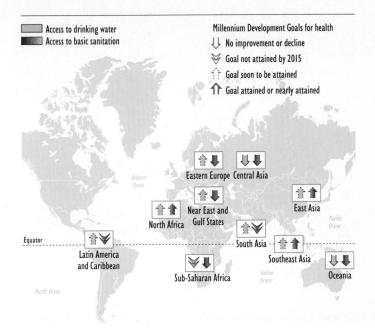

Figure 22 Interim progress made towards the Millennium Development
Goals for health in 2005: access to drinking water and access
to basic sanitation
From Atlas of Globalization, *Le Monde Diplomatique* (2006)

of 5.4 billion dollars. DATA does not include debt-relief aid in
its calculations.

DATA therefore urged German Chancellor Merkel to press for
an increase in development aid for 2007 of more than six billion
dollars at the G8 summit. Great Britain and Japan pledged the
biggest rises; the USA and Canada were in the middle field; while
Germany, France, and Italy trailed at some distance behind.

Needless to say, the healthcare sector is also affected by this shortfall. Although the USA in particular has stepped up efforts in the fight against AIDS and malaria, the goal of providing treatment for every HIV-positive individual remains a remote one.

Doctors Without Borders

Several developments have been set in motion in recent years. In view of the often considerable sums required, smaller organizations tend not to appear much in the limelight. However, a nongovernmental organization that repeatedly captures public attention is Médecins Sans Frontières (MSF), or, as it is known in the USA, Doctors Without Borders, which helps people in need the world over. MSF was founded in Paris in 1971. One of the founding members was Dr. Bernard Kouchner, who – despite being a member of the French Socialist Party – has served as the Minister of Foreign and European Affairs in the administration of the conservative French President Sarkozy since May 2007.

MSF was originally active in disaster areas and regions beset by armed conflict. Now, however, it is increasingly providing medical care to people afflicted by AIDS, tuberculosis, malaria, and neglected diseases. The organization has launched the DNDI (Drugs for Neglected Diseases Initiative), whose aim is to promote the development of urgently needed drugs for tropical diseases. Together with other organizations, the initiative is investing 250 million dollars in new drugs for sleeping sickness, Chagas disease, and kala-azar. MSF is increasingly committed to the abolition of patent rights for vital drugs and wants the big pharmaceutical companies to supply such drugs in the developing world at affordable prices. In 1999 MSF was awarded the Nobel Peace Prize in recognition of its work helping victims of acute crises and violence.

8.13 The specialists

UNAIDS is a United Nations program that focuses its efforts entirely on the HIV/AIDS crisis. It coordinates various programs for fighting the pandemic, gathers the latest epidemiologic data, and evaluates them in an economic and social context. In addition, UNAIDS mobilizes the political community, industry, and society. Its aim is to formulate a global strategy for controlling the disease.

The biggest state-run specialists are PEPFAR – the President's Emergency Plan for AIDS Relief, which is wholly funded by the US government, and the Anti-Malaria Initiative in Africa, which were set up in 2004 and 2005 and, with a budget of 15 billion and 1.2 billion dollars, respectively, provide substantial funding in selected countries. Shortly before the 2007 G8 summit, US President Bush pledged a further 30 billion dollars for the following five years as a continuation of the PEPFAR program, which was scheduled to end in 2008. During his Africa trip in early 2008, the President announced another 350 million dollar grant for the fight against neglected tropical diseases. Much criticism can be leveled at this president, but in the fight against HIV/AIDS and other health threats he is at the forefront. In July 2008 the PEPFAR budget was even further increased by the US Congress to 48 billion dollars – 39 billion for AIDS, five billion for malaria and four billion for tuberculosis.

8.14 Public-private partnerships (PPPs)

By the end of the 20th century many people had concluded that strategies for fighting diseases were obsolete and that new organizational forms were needed that respond flexibly and transparently, are better oriented to the needs of recipients, and at the same time make the latter aware of their obligations. Until then financial support had been paid directly to governments. Time and time again, though, the money did not end up helping the needy but instead landed in the accounts of corrupt rulers or was absorbed by bureaucracy. Other projects were never implemented. Many were simply not thought through. In many cases too much emphasis was placed on direct success. "Give a man a fish; you have fed him for today. Teach a man to fish; and you have fed him for a lifetime." This well-known proverb captures the essence of the dilemma we face. Conventional development aid all too often hands out fish and all too rarely fishing tackle.

GAVI

To counter these problems with regard to infectious diseases and especially to promote immunization, UNICEF, the WHO, industrial partners, the World Bank and private foundations joined forces in 2000 in the Global Alliance for Vaccines and Immunization (GAVI), a true PPP. GAVI remains a driving force behind the immunization of children in seventy-two countries.

Global Fund to Fight AIDS, Tuberculosis and Malaria

The Global Fund to Fight AIDS, Tuberculosis and Malaria, which was founded in the same year, works along similar lines. It is jointly funded by international and national organizations as well as by foundations and private industry. Its aim is to combat the major epidemics: AIDS, tuberculosis, and malaria. The Global Fund (like GAVI), is trying to turn the system upside down, i.e. by not telling countries what has to be done but rather being receptive to their wishes and needs. Groups within developing countries can apply for project funding. These are then examined with regard to their feasibility and importance. Approved projects are sponsored for years, and the funding is periodically reviewed. The applications must be transparently supported with minimum red tape. More than 99% of the Global Fund's income reaches the recipients in developing countries. Nearly half its money is granted to nongovernmental organizations.

To date the Global Fund has invested 11 billion dollars in programs to combat AIDS, tuberculosis, and malaria in 136 countries. It has provided 1.4 million people with ART for HIV/AIDS; 3.3 million people with tuberculosis drugs based on the DOTS strategy, and 46 million families with insecticide-treated mosquito nets. It has also distributed drugs that have prevented 23 million malaria cases. The Global Fund hopes to grant six to eight billion dollars annually by 2010. In their closing document in Heiligendamm, the G8 pledged six to eight billion dollars in further aid. At the subsequent conference in Berlin in late September 2007, seven billion euros was actually collected for the Global Fund, thus ensuring a firm financial basis for the Fund in the coming years.

8.15 Foundations

From computer billionaire to philanthropist

The organization that has had the greatest influence in bringing about the sea change in the fight against diseases in poor countries is a private foundation: the Bill & Melinda Gates Foundation. With an endowment of 30 billion dollars, which will increase steadily to 60 billion dollars over the next few years, it is by far the biggest foundation in the world. Its founders, Melinda and Bill Gates, became multibillionaires through Microsoft, the software company. In 2008 Bill Gates retired from the daily business of running Microsoft and became engaged full-time with the

Foundation. The Foundation was founded in 2000 with an initial endowment of 100 million dollars which steadily increased to 30 billion dollars. In 2006 that figure was doubled overnight when Warren Buffett, an investment banker, 75 years old at the time and one of the world's richest men, announced that he intended to donate his shares in the Berkshire Hathaway investment pool to five foundations. He bequeathed the lion's share, worth 30 billion dollars at the time, to the Gates Foundation (see box: "The biggest donation of all time"). Buffett's letter to Melinda and Bill Gates summarizes the conditions in just two pages and is worth around 60 million dollars per word. Of the Bill & Melinda Gates Foundation's three programs, two are international in scope, while the other is limited to the USA. The latter aims primarily to improve the catastrophic educational conditions for America's poorest. The Global Development Program donates money to, among other things, agricultural projects. The Global Health Program fights diseases where there is an especially crass inequality between rich and poor. Accordingly, the major infectious diseases and immunization programs to fight childhood diseases are priority areas. Thus the Foundation is strongly committed to the eradication of polio through immunizations; it is a major supporter of GAVI and the Children's Vaccine Program. The Foundation grants around one billion dollars a year to address health issues. Besides immunization programs, the Foundation has recently also started to look at longer-term approaches and is now supporting new lines of research.

In early 2007 a report in the *Los Angeles Times* (of January 7, 2007) cast doubt on the untarnished image of the super-foundation. It revealed that the Bill & Melinda Gates Foundation holds shares in a number of companies that earn money in business areas which the Foundation opposes. (For every dollar the Foundation spends on good works, it invests 19 dollars in increasing its endowment.) Thus, the Foundation invests more than 8.5 billion dollars, or 41% of its capital, in companies that are diametrically at odds with its own principles, including over 100 million dollars each in pharmaceutical multinationals such as Abbott Laboratories, Schering Plough Corporation, and Merck & Co. According to the report, the Foundation also has considerable holdings in the big oil companies ENI, Shell, and Exxon Mobil (ESSO), all of which are contributing to environmental pollution by flaring gas at oil plants in the Niger Delta, where numerous children are afflicted by respiratory diseases. Is it acceptable for children to be protected against infectious diseases on the one hand and then suffer from respiratory diseases on the other?

Of course money is invested in dubious companies elsewhere too. In many industrialized nations, for instance, the state's coffers are swelled by tax money from companies that are not entirely above reproach. Nevertheless, it is a fair question to ask if the world's biggest charity foundation should not assume a model role.

From pharmaceutical company to research trust

The Wellcome Trust, Europe's biggest medical research trust with net assets of 12 billion pounds sterling (18 billion euros), handles things differently. The Trust was set up in 1936 on the death of the pharmaceutical magnate Henry Wellcome. It is an independent organization that promotes biomedical research with the aim of improving the health of humans and animals. The trust honors its social principles not only in its investments but also exercises a strong social conscience at the general shareholders' meetings of those companies in which it has major holdings.

The biggest donation of all time

"Brace yourself," Warren Buffet is reported to have said to the editor of Fortune magazine, Carol J. Loomis, during a meeting in the spring of 2006. "I know what I want to do", he said, "and it makes sense to get going." Buffet, the CEO of Berkshire Hathaway Inc., was planning to make the biggest donation in history. On June 26, 2006 he put it all in writing. In a two-page letter, which was published, he donated 30 billion dollars of his fortune to the Bill and Melinda Gates Foundation (printed in full below):

Dear Bill and Melinda:

I greatly admire what the Bill & Melinda Gates Foundation ("BMG") is accomplishing and want to materially expand its future capabilities. Accordingly, by this letter, I am irrevocably committing to make annual gifts of Berkshire Hathaway "B" shares throughout my lifetime for the benefit of BMG. The first year's gift will permit an increase in BMG's annual giving of about $1.5 billion. In the future, I expect the value of my annual gifts to trend higher in an irregular but eventually substantial manner.

Here are the mechanics: Ten million B shares will be earmarked by me for BMG contributions. (I currently own only A shares but will soon convert a number of these to B.) In July of every year, or such later date as you elect, 5% of the balance of the earmarked shares will be contributed either directly to BMG or to a charitable intermediary that will hold the earmarked shares for the benefit of BMG. To illustrate, in 2006, 500,000 shares will be contributed. In 2007, 475,000 shares (5% of the 9,500,000 remaining after the 2006 contribution) will be contributed and thereafter 5% fewer shares will be contributed each year.

There are three conditions to this lifetime pledge. First, at least one of you must remain alive and active in the policy-setting and administration of BMG. Second, BMG (or any intermediary) must continue to satisfy legal requirements qualifying my gifts as charitable and not subject to gift or other taxes. And, finally, the value of my annual gift must be fully additive to the spending of at least 5% of the Foundation's net assets. I expect there to be a ramp-up period of two years during which this condition will not apply. But beginning in calendar 2009, BMG's annual giving must be at least equal to the value of my previous year's gift plus 5% of BMG's net assets. If this amount is exceeded in any year, however, the excess can be carried forward and be offset against a shortfall in subsequent years. Similarly a shortfall in a given year can be made up in the following year.

The value of Berkshire shares will, of course, vary from year to year. And, as noted, the number of shares distributed will diminish by 5% per year. Nevertheless, I believe that you can reasonably expect the value of Berkshire shares to increase, in an irregular manner, by an amount that more than compensates for the decline in the number of shares that will be distributed.

BMG can rely on this pledge to immediately and permanently expand its activities. My doctor tells me that I am in excellent health, and I certainly feel that I am. If I should become incapacitated, however, and be unable to administer my affairs, I direct whoever is in charge of my affairs to honor the commitment I am making in this letter. Additionally, I will soon write a new will that will provide for a continuance of this commitment – by distribution of the remaining earmarked shares or in some other manner – after my death.

I regard Berkshire as an ideal asset to underpin the long-term well-being of a foundation. The company has a multitude of diversified and powerful streams of earnings, Gibraltar-like financial strength, and a deeply-imbedded culture of acting in the best interests of shareholders. Outstanding managers are available to succeed me. I expect Berkshire to become ever-stronger and more profitable as it makes new acquisitions and expands present businesses.

I hope that the expansion of BMG's giving is one of depth, rather than breadth. You have committed yourselves to a few extraordinarily important but underfunded issues, a policy that I believe offers the highest probability of your achieving goals of great consequence. The doubling of BMG's present spending can increase the Foundation's already impressive effectiveness in addressing the societal problems upon which it now focuses.

Working through the Foundation, both of you have applied truly unusual intelligence, energy and heart to improving the lives of millions of fellow humans who have not been as lucky as the three of us. You have done this without regard to color, gender, religion or geography. I am delighted to add to the resources with which you carry on this work.

Sincerely,

Warren E. Buffett

Original on the internet at: http://berkshirehathaway.com/donate/bmgfltr.pdf

9 Swimming Against the Tide

The future depends on what we do in the present.

Mahatma Gandhi

In the last quarter of the 20th century pharmaceutical companies launched nearly 1400 new drugs on the market, more than fifty a year. Of those, about 180 were for the treatment of cardiovascular diseases, but only three for tuberculosis, four for malaria, and just thirteen for all neglected tropical diseases put together. Looking at the market for antibacterial drugs, we see a downward trend: today only around a dozen pharmaceutical companies are producing such drugs, compared to around seventy as recently as the 1990s. In the ten years between 1983 and 1992 they introduced about twenty new antibiotics, in the ten subsequent years seventeen, and in the four years since 2003 just four. What it all boils down to is this: an estimated 100 billion dollars is currently being invested in research and development in the healthcare sector every year. Of this, only one tenth goes towards diseases that affect developing countries, i.e. infectious diseases, which make up 90% of the global disease burden. The situation will remain unchanged as long as economic considerations play a decisive role in the development of new drugs.

9.1 The quest for blockbusters

What's wrong? With annual sales topping 600 billion dollars, the global pharmaceutical market is one of the most attractive industrial sectors in existence. Pharmaceutical companies are three times more profitable than the 500 biggest nonpharmaceutical companies put together. On average they generate a 15% profit. The Fortune 500, the 500 biggest companies in the USA, include ten pharmaceutical companies, which earn more than the other 490 companies combined.

Blockbusters, or drugs with sales in excess of one billion dollars a year, are especially attractive to these companies and account for more than a third of all drug sales. In 2005, ninety-four drugs were rated as blockbusters. Most are drugs for the treatment of diseases that affect a large number of people chronically or drugs that represent a breakthrough in treatment or prevention. Many of the pills, drops, and capsules only alleviate the symptoms – not the causes – of a disease. They are usually prescribed to a large group of people over a long period of time. They include statins, drugs for the prevention of cardiovascular disease. In many cases physicians prescribe statins preventively; in Great Britain there was even a proposal to add them to the drinking water.

In 2005 the leading blockbuster was the statin Lipitor®. This drug alone poured more than 12 billion dollars into the coffers of Pfizer. It was followed by Advair Diskus® (GlaxoSmithKline) for the treatment of asthma and other pulmonary conditions with sales of around 5.5 billion dollars, the gastric acid blocker Nexium® from AstraZeneca with sales of nearly 5 billion dollars, and the statin Zocor® from Merck with sales of 4.4 billion dollars. Other products, such as the antidepressants Prozac® and Zoloft®, have taken on the character of "lifestyle drugs," i.e. they enhance a patient's feeling of well-being.

However, the whole business is exceedingly short-lived: block-busters need to bring in the cash quickly. Often the exclusivity rights are limited to just one or two years. Once the patent expires and the substance is free for generic manufacturers to copy, prices plummet. For example, while it was under patent protection, each capsule of the antidepressant Prozac cost around 2.50 dollars. Since then the price has fallen to one tenth of that amount.

Of course these drugs have improved or saved untold numbers of lives. But without exception they are products intended for industrialized countries. Ninety percent of all drugs manufactured are consumed by just one-seventh of the world's population. Over half are sold across pharmacy counters in the USA and Canada alone. The interests of economists and physicians usually coincide only where chronic diseases affecting wealthy nations are concerned.

9.2 Economic viability

The situation is completely different in the case of diseases that are rare or nonexistent in wealthy countries but pose a major problem in the developing world. "Blockbuster" is not a term that is easily applied to antibiotics and vaccines. The markets for these products are small. In 2005 the vaccine market was just 10 billion dollars, less than the market for the statin Lipitor® alone. Only a single new anti-infective agent is expected to take off in a big way, namely the vaccine against human papillomavirus (HPV), a cause of cervical cancer. The reason for this is that routine vaccination against HPV is now recommended for all girls – at least for all girls in many industrialized countries, and because the required three injections cost 500 euros. If the cost remains this high, the incidence of the second most common form of cancer in women is unlikely to change much on the world stage, because 80% of the women at risk live in poor countries.

When assessing research-and-development projects that address infectious diseases, it has proved helpful to distinguish between three types of diseases. Type 1 are those that are relevant to industrialized countries and therefore more or less fit into their research-and-development portfolio. They include hepatitis B and HVP-induced cervical cancer. Type 2 diseases also occur in industrialized countries but are far more common in developing countries, for example HIV/AIDS and tuberculosis, and are still of some interest to the pharmaceutical industry. Type 3 diseases largely affect only poor countries, for example the neglected tropical diseases. Almost nothing is being invested in this area. Profit expectations and urgency of need are diametrically opposed concepts.

9.3 No one needs flops

A drug costs money long before it is marketed. Most drug candidates fall by the wayside along the path to approval. The pharmaceutical industry calculates that it invests 800 million dollars in every drug that is approved before it reaches the first real patient. Although one can argue about how realistic this figure is, its magnitude makes it clear that strict selection criteria have to be applied. For type 2 drugs the decision as to whether to pursue development may or may not be positive. Indeed, the HIV drug AZT (azidothymidine, Retrovir®) has emerged as a blockbuster and makes up one component of the ART regimen. Yet originally AZT was developed as a cancer drug in the 1960s. In this case the research-and-development costs have long since been recovered, and the drug should therefore be made available in developing countries at an affordable price.

In the case of type 3 drugs, by contrast, the gap between financial incentive and medical need inevitably leads to a dilemma. Discussions about withdrawing the patents for certain drugs only serve to frighten the pharmaceutical industry more, with the result that some type 2 drugs could be reclassified as type 3. That has already happened with tuberculosis drugs. Tuberculosis drugs are all over thirty years old and have long since lost their patent protection. Very few companies find it interesting to manufacture these drugs. If their prices fall further, supply bottlenecks are likely to occur. The development of new tuberculosis drugs has stalled completely. The outcome is that no effective drugs are available today for XDR-TB.

The vaccine market is even trickier. Here the pre-approval hurdles that have to be cleared are even higher. Moreover, the profit margins for vaccines are small. Also, research and production are different in the vaccine market. Unlike the small synthetically manufactured molecules of typical pharmaceutical products, the components of vaccines are living or inactivated microbes or genetically engineered parts of the microbe. Only half a dozen or so companies in the world have the requisite expertise. They are nearly all located in western Europe and the USA, and their production output only just meets current needs. If there were a sudden surge in demand, shortages could occur.

9.4 Talking doesn't help

We need incentives for developing new anti-infective agents. To this end, synergies between researchers in the public and private sectors have to be strengthened. The development of anti-

infectives is best promoted through targeted incentives. Changes
to patent laws are also called for.

The relevance patent

According to Article 8 of the TRIPS Agreement, countries can
adopt coercive licensing practices in order to maintain the health
of their population. There is little incentive to develop drugs to
which this clause is likely to apply. Languishing in the drawers
of pharmaceutical companies are many starting substances
for potential anti-infectives whose further development is not
deemed worth the effort. There have been many proposals for
changing patent laws. One could, for example, base the granting of patents not just on the novelty of a product but also on
its importance to global health. It would then be possible to
introduce a second patent model. Drugs and vaccines could be
assessed according to their relevance (measured in deaths and
DALYs) and an imbalance factor (calculated from the unequal
significance of a disease in developing and industrialized countries). Depending on the benefit of the drug for humanity, the
manufacturer would receive payment in exchange for waiving
exclusive rights to manufacture it. The payments would be linked
to progress, with partial payments being made at the time of
patent registration, at the start of clinical trials, and at the time
of licensing. The money could be provided from a fund set up on
the PPP model, which is financed jointly by the public sector, the
pharmaceutical industry, and foundations.

The potential advantages are obvious. Pharmaceutical companies and, even more so, startup biotech companies would devote
greater energy to developing these drugs, as a certain profit would
be guaranteed. The staged payment, in turn, would ensure that
the procedure is followed to the end. Finally, the products could
be sold at affordable prices. The approach could be refined so

that a pharmaceutical company could revert to the conventional patent procedure at any time provided that it was prepared to price the product at an affordable level in developing countries. Needless to say, any payments already made to the pharmaceutical company would then have to be refunded.

Once the poverty-associated diseases have been weighted according to their relevance and imbalance, the same factors could also be applied to the disbursement of funds for research programs. I'll return to this point later.

Vital assistance for chemical orphans

A banner that states: "A message from research-based pharmaceutical manufacturers: we don't neglect any disease, even if it only affects 0.00625% of the population." This is just one of many slogans used in an ambitious three-year, 30-million-euro campaign launched by the Association of Research-Based Pharmaceutical Companies in Germany. The aim is to polish the industry's tarnished image. These powerful lobbyists are also launching numerous activities to address the patent issue. The 0.00625% figure refers specifically to the prevalence of pulmonary hypertension in Germany. The Association of Research-Based Pharmaceutical Companies in Germany is using this example of a disease relevant to the rich countries to explain the principle of special regulations that apply to drugs for rare diseases, known as orphan drugs.

The government has created incentives for the development and marketing of such drugs which, according to sober economic calculations, would not be financially worthwhile. In the USA the development of drugs and vaccines against diseases that affect fewer than 200,000 Americans has been promoted through tax-relief measures since 1983. Companies are granted a guaranteed monopoly on the drug for seven years. The European Union passed a similar directive in December 1999, which grants exclusive sales rights for up to ten years on drugs and vaccines for diseases affecting fewer than one in 2000 people. The regulations also apply, irrespective of case numbers, to drugs which a committee decides would be unlikely to be marketed without incentives. In addition, investors can be exempted from licensing fees levied by the European Medicines Evaluation Agency (EMEA).

Sometimes, evaluations have proven wrong, and apparent "orphans" have turned out to be blockbusters. For example, in 1989 the US company Amgen was given the green light to produce Epogen® under the provisions of the Orphan Drug Act. The drug, popularly known as "epo" (erythropoietin), is a hormone that stimulates the production of red blood cells. It has come under the public gaze in recent years as a doping agent used by competitive cyclists. Originally Amgen intended to market the drug for the treatment of anemia in end-stage renal failure. However, it soon became clear that epo is also beneficial to patients with bone-marrow dysfunction, e.g. those taking AIDS drugs or chemotherapy drugs for cancer. In 2001 Epogen® and another epo product had climbed to sixth and seventh places on the US bestseller list of drugs. Epogen® alone generated sales of 2.4 billion dollars in 2003.

According to the US Federal Drug Agency (FDA), since the Orphan Drug Act came into force in 1983, 200 drugs for rare diseases have been marketed. In the preceding decade fewer than ten had. It is therefore worth considering introducing a similar regulation for widespread but neglected diseases. Orphan drugs would then be defined not only on the basis of the rarity of the disease for which they are intended but also on the basis of weak interest in them in industrialized countries. In this case the exclusivity right would probably be a hindrance and would have to be compensated for by other incentives.

Recently, an interesting proposal came up in the USA. A transferable review voucher for every drug approved of a neglected disease is being offered. This voucher guarantees accelerated review by the FDA (from 8–12 months down to 6 months). The voucher is valid for any kind of drug and can even be sold to another company. Imagine such a voucher being offered on eBay! It could be worth up to several hundred million dollars if used for a drug that then becomes a blockbuster.

9.5 New incentives

Regulated dual-price system: The dual-price system, according to which the same drug is sold at a high price in industrialized countries and at a low price in developing countries, has often been criticized as being at odds with the free market principle. But charging different prices for the same services has long been reality in many areas. In fact, the only price asked for is the one that is affordable.

When a manufacturer forgoes profits in developing countries and grants production rights to a generics manufacturer (possibly without claiming patent rights and licensing fees), the price of a drug can easily fall by an order of magnitude (see box "Copycats").

Opponents of this scheme often point to the danger of reimports as an argument. But steps could be taken to counter this by coding packages, dying the drugs, and so forth. At any rate, it is clear that recipients in developing countries must be integrated in the control measures and attendant obligations.

Purchase guarantee at a fixed price: Under this scheme international organizations and foundations would guarantee a future sales market while a drug or vaccine is still under development

Copycats

A generic is a drug that has more or less the same spectrum of activity as another drug that already exists under a proprietary name. It may differ from the original product in their formulation and dosage form. Many companies manufacture generics, which are gaining importance as developing countries increasingly invoke Article 8 of the TRIPS Agreement and adopt coercive licensing. In many cases generics companies not only produce drugs more cheaply but also in better formulations. They can, for example, formulate combination products using drugs from different manufacturers. But there are also black sheep among them who imitate drugs illegally, counterfeiting products that are still under patent protection, and often in inferior quality. A WHO study came to the conclusion that up to 50% of all counterfeit drugs are totally ineffectual because they contain no active substance. In 17% of counterfeit products the wrong drug was used, and in one in ten the content of the drug was insufficient. Catastrophic conditions in plants producing cheap drugs have been uncovered in China. In 2005 alone well over 100,000 unlicensed small manufacturers and over 400 illegal drug production sites were closed down. In just three months in 2006, 14,000 drugs were newly licensed. Something was not quite right. In fact, in 2006 at least ten patients died as the result of the side effects of a contaminated antibiotic. In mid-2007 the Chinese government took action. The Director and General Secretary of the Chinese regulatory agency for food and drugs were condemned to death and several of their coworkers received long prison sentences for corruption. The Director had allegedly collected over 800,000 dollars in bribes. The General Secretary had approved over 270 drugs in return for bribes, at least six of which were ineffective imitations.

and commit themselves to purchase a quantity of the product at an agreed price.

This scheme has been practised in the past. Thus, the British government, in an endeavor to advance the development of a meningococcal vaccine, guaranteed sales to the first manufacturer that could come up with such a vaccine. The governments of the USA and Germany amongst others paid for smallpox vaccines in order to be armed against the possible threat of a bioterrorist attack and did so – if not before development – at least prior to production. What company would otherwise bother to produce a vaccine for a disease that has already been eradicated?

In 2007 the United Kingdom, Italy, Canada, Norway, and Russia together with the Bill and Melinda Gates Foundation committed 1.5 billion dollars for the development of a new pneumococcal vaccine. The advanced market commitment (AMC) of these donors aims to create a market for the vaccine in developing nations at a guaranteed fixed price, for a defined amount of doses, and for a certain number of years and requires that the manufacturer then supplies the vaccine at the prefixed price. It is my hope that the AMC scheme will be used more often in the future.

Targeted tax relief for pharmaceutical companies is also a conceivable approach and is already being practiced in some countries. Of course, it must be ensured that the money saved is actually invested in the intended projects. It is helpful when the company concerned sets up independent research institutes based on the foundation model. The company's research laboratories then collaborate with those institutes, e.g. by supplying lead substances. In addition, cooperation between the institute and scientists and clinicians at public institutions is facilitated. At the same time there is a clearcut separation between the foundation-like research institute and the sponsoring pharmaceutical company. The Swiss pharmaceutical giant Novartis has already implemented this scheme. In 2002 it set up the Novartis Institute for Tropical Diseases in Singapore together with the Singapore government, where it is developing drugs for tuberculosis and dengue fever. The concept appears to have met Novartis's expectations, since the company is now setting up the Novartis Vaccine Institute for Global Health in Siena, Italy based on the same model. Its aim is to develop vaccines against poverty-related infectious diseases with emphasis on diarrheal diseases.

Finally, the *trading of debts* in healthcare programs in developing countries should be considered. In this scheme developing countries would be granted debt relief if they agree to invest some of the debt in healthcare programs. This would specifically call upon the services of the World Bank, the International Monetary Fund, and rich donor countries. We will return to this option in Chapter 11.

9.6 Pooling expertise

Partnerships between public sector research institutes and private industry could significantly reduce the investments needed to develop anti-infectives. Such PPPs are already being specifically supported through sponsorship programs such as those set up by the European Union, the Bill & Melinda Gates Foundation, the Wellcome Trust, and the National Institutes of Health in the USA. Each partner does what it does best. The public institutes carry out the basic research and preclinical studies, while the private sector is responsible for the clinical trials and product development. During the planning phase, agreements must be reached on patents and licenses and sales channels in developing countries. If possible the pharmaceuticals are pitched at an affordable price with the help of generics producers. This is the

In 1900 the German mathematician David Hilbert invited the community of mathematicians to take stock of their discipline. If you want to know where mathematics is heading, he said, you have to be aware of the current problems that will have to be solved in the future. Hilbert formulated twenty-three key problems, which he challenged his fellow mathematicians to address. Eighteen of those have now been solved. Only two problems remain unsolved; one was formulated too vaguely; and two others have been partly solved. Bill and Melinda Gates took inspiration from Hilbert's approach when they set up the Grand Challenges in Global Health Initiative. They asked scientists to list the most important health issues arising from inequality in the world. One thousand five hundred experts from around the world responded. A scientific committee reviewed the proposals and compiled a priority list of the most important problems facing world health. Many of the questions related to the major infectious diseases, their prevention by vaccines, new treatments, and early detection. An appeal was then launched to submit projects aimed at finding solutions. Five hundred researchers responded. Forty-five projects were approved and received grant money. The initiative aims to promote innovative approaches that will be brought to bear where they're needed most.

In October 2007 the Foundation announced a new innovation program worth 100 million dollars. It aims to sponsor 1,000 projects to the tune of 100,000 dollars each. Unusually, the applications only have to be two pages long and are immediately approved or rejected. No preliminary work is required. This revolutionary approach is meant to encourage young scientists in developing and threshold countries to give their imagination free rein and to come up with fresh, bold proposals which can then be tested experimentally with funding from the foundation.

approach taken by the Bill & Melinda Gates Foundation's global access strategy in the Grand Challenges in Global Health Initiative (see box: "Motivation to Ponder"). Patent and license fees for new products will be waived in developing countries but may still be charged on sales in industrialized countries.

PPPs already have a budget of more than one billion dollars for developing new drugs against the major infectious diseases. Many of those are substances patented by basic researchers at public research institutes who have agreed to allow any drugs derived from them to be sold in poor countries free of royalties. Pharmaceutical companies have also opened up huge chemical libraries to facilitate the search for new drugs. Thus, nearly forty new candidate drugs for type 2 and type 3 diseases have been developed in recent years.

9.7 A global fund as a clarion call

In its commentary on the Millennium Development Goals (see Chapter 8), the WHO Commission for Macroeconomics and Health issued proposals for sponsoring research into AIDS, tuberculosis, malaria, and neglected diseases. It proposed a fund for global health research which, modeled on the Global Fund to Fight AIDS, Tuberculosis and Malaria, would have a budget of one billion dollars in 2007, rising to two billion in 2015. Such a fund, in conjunction with other programs, could help focus research efforts on the health problems of the poor, thus rousing anti-infective drug development from its slumber. An important aim here is to promote and expand research capacities in developing countries and stimulate an exchange of researchers between industrialized and developing countries.

9.8 Push or pull to success

Most granting schemes follow the "push principle". An applicant formulates projects and goals and receives grant money to pursue them. In this way researchers have the freedom to choose what they believe is the right way forward. However, the sponsor bears the entire risk of failure. Push programs certainly remain the most important sponsorship instrument in basic research. But they need improvement. Many basic researchers in the biomedical field tend to seek out the easiest experimental model that is tailored to their questions and is able to provide answers within a reasonable time. This has proven effective for many basic questions. But a disadvantage is that the answers are not readily transferable; they only pertain to the model used. The drive for quick success with the help of the push principle disadvantages more complex science in the social context, i.e. investigations that directly address a socially relevant problem.

In the case of health issues that are relevant to industrialized countries, this drawback is avoided through specially tailored promotion programs for public sector institutes and the pharmaceutical industry. Things are different, however, in the case of poverty-related diseases, in which the industry has little interest and the public sector sees no relevance to its own society. Worldwide, less than 5% of research money is granted for investigating these questions. To remedy this, research grant applications could be weighted with a relevance factor and assessed on that basis (similar to the method proposed for granting patents).

In addition, pull programs should help overcome the current research impasse, because funding will be based on success. The financial backer sets up a kind of competition for research into a problem or for the development of a process or a product. The first one to deliver the process or a solution to the problem wins the grant. Here too funds can be released in stages, so that pull programs pay for success in increments. This means that competitors have to finance the groundwork themselves. Thus, absolute pull programs rule out basic research in the biomedical field, because advance financing is almost always required. But from the point where funding goes into clinical trials, pull programs can be extremely useful. Essentially, the Orphan Drug Act is an example of a pull program. The guaranteed purchase of a drug can also be seen as a form of pull scheme.

Quite apart from such procedural decisions, however, the key to success is to define funding priorities. There is a marked gradient even in the allocation of funds for projects concerning the major infectious diseases AIDS, tuberculosis, and malaria. Thus, in 2001 and 2002 more than two billion dollars flowed into projects for fighting HIV/AIDS, while only a fifth of that was spent on tuberculosis and an even smaller sum on malaria. The trend can be seen even more starkly at the US National Institutes of Health (NIH), the world's biggest research organization for medical sciences: In 2005 the NIH spent nearly three billion dollars on HIV/AIDS, only 158 million dollars of that on tuberculosis, and just 90 million dollars on malaria. The potential bioterrorism weapons smallpox and anthrax ranked ahead of tuberculosis and far outstripped malaria in funding. Neither the NIH nor the EU funds projects for, or research into, poverty-related diseases according to the degree of their societal importance.

9.9 Research incubators

To strengthen PPPs, flexible structures need to be implemented that work according to private sector principles and can be used jointly by cooperation partners in basic research, startup companies, and pharmaceutical multinationals. These "research incubators" can serve as nodes at which scientists from various disciplines can conduct research jointly in the field of major infectious and neglected diseases. Their main tasks should be the discovery and preclinical testing of anti-infectives and the development of new diagnostic agents. They could be supported by larger service units, which could offer, for example, broad screening of chemical libraries as well as global genomic and functional analyses.

This approach would not only streamline research work but would also facilitate communication between scientists working in different fields. Finally, such institutions could address questions relating to innovative combination therapies at an early stage, e.g. interventions comprising both vaccines and drugs. Ideally, they would also deal with issues such as intellectual property rights, licensing, global access, socioeconomic consequences, and ethics.

The financial starting capital for these incubators should be provided jointly by the public sector, pharmaceutical companies, and foundations. This would be supplemented by project-specific payments by users. In the case of researchers from public institutions, the work could be financed through individual public-sector support programs; in the case of private sector research work by the industrial partners. The research findings would be provided to developing countries without patent and license claims, and revenues would stem solely from sales in industrialized countries. The transfer of science, technology, and innovation between industrialized and developing countries would also be an essential element. Partner institutes in developing countries should be integrated in such a network. This is especially important in the case of sub-Saharan Africa, a region which, on a per capita basis, accounts for just 5% of all patents in developing countries and where there are fewer than twenty scientists and engineers per million inhabitants. Especially in the biomedical sciences, the African continent is suffering from a brain drain: up to half of all medical students who complete their training in South Africa work abroad. In Zimbabwe that figure is as high as two-thirds.

9.10 In the trenches

In order to make the translation from the laboratory to the field as smooth as possible, it will be important to develop capacities for clinical research and medical trials in developing countries. It is widely agreed that drugs and vaccines should first be tested for safety in humans in the country or region in which they were discovered. Only then should they be clinically tested to the same quality standards in countries where the disease is rampant. Few institutions in developing countries can do this at present. Creating robust capacities there will require start-up finance. The funding must benefit a large number of centers. Though top-drawer institutes are needed, biomedical and clinical research must also be strengthened at many locations. Otherwise a new dual-class system would emerge in developing countries. There must be no compromises with regard to the quality of drug trials in developing countries, i.e. the same high standards must be applied as in industrialized countries. Drugs for diseases that are equally prevalent in industrialized and developing countries, e.g. chronic cardiovascular diseases, should be tested in industrial-ized countries only. Centers in the poorest countries can then focus on those diseases that are endemic in, or unique to, those regions and for which there is a great deal of catching up to do.

A delicate question regarding the licensing of new vaccines

In 1998, soon after the introduction of the vaccine against rotavirus, a common cause of diarrhea in children, some US physicians observed intussusception, the enfolding of one segment of the intestine into another, in some immunized children. A link between vaccination and the disease was not proven, but the possibility could not be ruled out. The vaccine was withdrawn from the market in 1999. This was a completely understandable decision on the part of public health officials in the USA, where rotavirus infections are rarely fatal. But the setback also delayed the introduction of the vaccine in developing countries, where rotaviruses claim the lives of up to one million children a year.

This example illustrates a dilemma. It poses the question of whether the same standards should be applied in countries where a disease is rampant as in countries where it is rare. Should different vaccine licensing criteria be established in different countries?

As far as rotavirus immunization is concerned, the question has been answered. Two new vaccines devoid of the aforementioned side effect have been approved since 2006. But the question remains and will keep recurring – notably with regard to AIDS and tuberculosis vaccines. It is clear that even mild side effects are grounds for exclusion in countries in which very few people are affected by a disease. But what about countries in which one in three or four people is HIV-positive and six in 1000 have tuberculosis? Should vaccines be denied approval in these countries because of side effects which, in the given context, may be acceptable? At some point an answer will be needed and it is unlikely to be forthcoming until then.

This is especially true of vaccines and the quest for drugs that confer protection against naturally acquired infectious diseases.

With regard to the creation of top-flight clinical trial centers, the informed consent of participants is an important consideration. (See box: "A delicate question regarding the licensing of new vaccines"). The population of the country in which a drug is tested should later have access to it – if possible free of charge or at least at an affordable price. Often there is little trust in Western science or confidence that the drugs, once developed, will actually be made available. Ultimately, how things work out also depends on the political leadership in each country. Whereas South Africa's first freely elected president, Nelson Mandela, vigorously took up the fight against HIV/AIDS and tuberculosis, his successor Thabo Mbeki has long denied that HIV is even the cause of AIDS. The minister for health recommended that vitamins and mixtures of raw herbs, vegetables and fruits should be taken to treat AIDS, and the vice-president believed that showering after sexual intercourse is enough to prevent AIDS transmission. A similarly abstruse spectacle is unfolding in The Gambia.

10 Hot Spots for Old and New Epidemics

A dried up tree does not refuse to burn.

From the Congo

10.1 Introduction

The subject of infectious diseases is and will probably always remain an open chapter. At least once a year an epidemic breaks out somewhere in the world. Some are due to familiar pathogens with altered characteristics. Some are caused by existing but previously undetected microbes that suddenly gain the upper hand. And some are caused by completely new agents. In the past thirty-five years more than thirty new microorganisms have been discovered that are potentially dangerous to humans. The list includes rotavirus (discovered in 1973), Ebola virus (1977), *Legionella* (Legionnaire's disease, 1977), *Campylobacter jejuni* (severe diarrhea, 1977), *Borrelia burgdorferi* (Lyme borreliosis, 1982), HIV (1983), hepatitis C virus (1989), Nipah virus (encephalitis or meningitis, 1999), and the SARS virus (2002).

The outbreak of SARS and type H5N1 bird flu have not only rapidly damaged the global economy, they have also once again raised the specter of new pandemics. Although new infectious diseases could flare up anywhere in the world, there are hot spots that favor their emergence and spread.

Eight factors that favor epidemics

- Infectious diseases have often been declared dead, but their threat remains because the necessary alertness is lacking.
- Existing epidemics will not be controlled as long as poverty prevails. They will continue to threaten not only developing countries but also industrialized countries.
- Catastrophes and conflicts exacerbate the situation, especially when refugees are penned up in unhygienic conditions.
- Fanatics need no deep understanding of microbiology and gene technology to use existing pathogens as weapons for terrorist attacks.
- Global warming especially favors the spread of vector-borne and diarrheal diseases.
- Exploitation of wilderness due to increasing contact between humans and wild animals as well as industrialized livestock production with increasing contact between humans and domestic animals, are dangerous breeding grounds for new pathogens that can potentially cause epidemics.
- The improper use of antibiotics accelerates the development of resistant microbes.
- Never before has it been so easy for an epidemic to spiral into a pandemic. Diseases are carried beyond national borders and across continents by refugees and travelers as well as by the global trade in animals and meat products.

10.2 Hot spot number 1: poor and sick, sick and poor

"There was a period in development thinking – not so long ago – when spending on public services, such as health and education, would have to wait. Good health was a luxury, only to be achieved when countries had developed a particular level of physical infrastructure and established a certain economic strength. [...] Experience and research over the past few years have shown that such thinking was at best simplistic, and at worst plainly wrong. [...] In my judgment, good health is not only an important concern for individuals, it plays a central role in achieving sustainable economic growth and an effective use of resources." Thus commented the former Director General of the WHO, Dr. Gro Harlem Brundtland. Improved health can break the vicious circle. Healthy, well-nourished people can devote more energy to their livelihood and start improving their own – and their country's – economic situation. The AIDS crisis shows the downward spiral very clearly: in many countries the productive population is draining away. In Botswana, one of the worst affected countries, three to four people are hired for every skilled job. For families, the disease means a huge increase in household expenditures and at the same time a fall in income. Children have to pitch in, especially with farming chores. Schooling becomes a luxury. Family units care for orphans and half-orphans as well as sick relatives and consequently are also unable to work. In South Africa many companies are already limiting the conditions for release from work, and only those who have lost a spouse, parent, or child are allowed to take time off. The vicious circle continues to turn: those who work less set aside less money for old age. For the children of these families, poor education lays the foundation for a life of penury.

In two virtual countries that are socioeconomically identical and differ only in that the life expectancy is five years longer in one, this factor alone results in an 0.3 to 0.5% higher annual economic growth rate. The economic growth of a country depends directly on the health of the population. Moreover, longer life expectancy acts as a stimulus for better education and more savings, both of which tend to strengthen the economy.

It is sometimes claimed that major epidemics act as a corrective factor for population growth in poor countries. This assertion is not just cynical but nonsense. Quite the opposite is true. There exists a well documented correlation between poverty and birth rates. The smaller the chance parents have for their newborn to survive, the more children they bring into the world in order to ensure that they can be cared for by their own children in old age. With a mortality rate of 5% for newborn babies, 15% for young children, and 10% for five-year-olds, only one in three children survives. To be on the safe side, poor parents reason, they must have five or more children. In sub-Saharan Africa each woman bears on average 5.5 children. It is hardly surprising that this also increases the mortality risk for expectant mothers.

The situation is becoming acute. In sub-Saharan Africa life expectancy has fallen dramatically in recent years, largely as a result of the AIDS crisis. Whereas life expectancy in South Africa was sixty years in 1995, by 2004 it had fallen to forty-eight. In Botswana it has plummeted by nearly 50%. A child born today in Zambia, Zimbabwe, Sierra Leone, Malawi, or Mozambique can expect to live half as long as a child born in Germany. Shortly before the dramatic downward turn, growing prosperity and improved healthcare systems had raised life expectancy to over sixty years in developing countries. All the high hopes those improvements engendered have been cruelly dashed. Life expectancy in forty countries will continue to fall until 2010.

Another vicious circle is also at work: The worse the economic situation of a country, the less is spent on healthcare and the greater the burden on individuals. Poor countries spend around 23 dollars annually on healthcare. In threshold countries that figure is on average five times greater and in industrialized countries one hundred times greater. In Africa fewer than 8% of the population have health insurance, and most policies do not cover prescription drugs for outpatients.

10.3 Hot spot number 2: catastrophes, conflicts and the threat of epidemics

Forty million people in the world are displaced and 10 to 20 million people are refugees. Sub-Saharan Africa alone has four million displaced persons and refugees, three-quarters of them women and children. Most are languishing in overcrowded camps, where tuberculosis and other respiratory diseases are rife. Where adequate sanitation systems and drinking water are lacking, diarrheal diseases, especially cholera and rotavirus infections, are rampant. Skin infections are also common. Often a single index case is sufficient to trigger an outbreak. In 1997 cutaneous leishmaniosis broke out in a refugee camp for Afghans in Pakistan. A single individual had brought in the disease, which is transmitted by sand flies. Subsequently, one-third of the people in the camp contracted the disease. Mosquito nets are almost never provided in refugee camps.

Girls and women are especially at risk in crisis situations. All too often they are subjected to sexual abuse. Hunger and desperation drive many into prostitution. Sexually transmissible diseases, especially AIDS, then spread rapidly. This is not restricted to crises. A recent study in the US city St. Louis found that more than a quarter of female drug addicts – one of the highest risk groups for HIV – had experienced enforced sex by policemen.

When the earth quakes and storms rage

The United Nations declared the last decade of the 20th century to be the International Decade for Natural Disaster Reduction. Unfortunately this did nothing to help. The decade witnessed more, rather than fewer, disasters. At the end of the decade, more than 80 major disasters had caused losses amounting to 600 billion dollars. In 2006 alone the WHO counted over 350 natural disasters in more than 160 countries, which blighted the lives of over 100 million people and drove half a billion people from their homes. Here are two examples that are still fresh in many people's minds:

On December 26, 2004 a tsunami devastated parts of Southeast Asia. More than 230,000 people died, and millions were made homeless. The countries hit hardest were Sumatra, Sri Lanka, India, Thailand, Myanmar, Malaysia, Indonesia, and Bangladesh. The tsunami reached as far as the eastern coast of Africa. Diarrheal diseases, especially typhoid fever and cholera, raged in many of the devastated areas. To counter the threat of epidemics, India launched a large-scale vaccination campaign against typhoid fever and cholera. As the water receded, it collected in ponds and puddles. These became breeding sites for mosquitoes, resulting in outbreaks of malaria and dengue fever in the crisis areas. Thankfully, large-scale epidemics did not materialize.

In August 2005 Hurricane Katrina destroyed large swaths along the Gulf Coast of the USA. Katrina ranked not only as the most destructive natural disaster ever to hit the USA but also as the costliest, causing 81 billion dollars worth of damage. Some 1800 lives were lost. Diarrheal diseases caused by shigellae, salmonellae, and noroviruses as well as cholera flared up. However, only one norovirus outbreak in Texas took on major proportions. New Orleans, which was hardest hit, was home to many people with HIV/AIDS or tuberculosis. The hurricane cut off many of them from their supply of medications. Around 8000 people with HIV/AIDS did not receive any medications for a considerable time and consequently suffered from severe symptoms of the disease.

In 2008 two major catastrophes hit Asia: Hurricane Nargis killed approximately 100,000 people in Myanmar, and an earthquake in Sichuan, China claimed up to 80,000 lives. In Myanmar support from outside was refused, thus significantly increasing the risk of outbreaks of infectious diseases such as cholera.

Natural disasters carry an increased risk of infection, especially for diarrheal and respiratory diseases. However, in many cases rapidly deployed international relief missions are able to contain the risk of major epidemics.

In the firing line

Depending on the calculation, experts reckon that 20 to 30 armed conflicts are currently being waged in the world. Since 1990 armed conflicts have cost Africa a total of 300 billion dollars, or 18 billion dollars a year. If one adds to these costs damage and loss caused by armed criminal actions and other acts of violence, 20 billion dollars a year is lost that could be used to deal with more pressing issues such as the Millennium Development Goals in Africa. In the past ten years military conflicts have claimed the lives of two million children. They have made five million children invalids, and 12 million homeless. More than one million children have been orphaned or separated from their parents, and untold millions have been permanently traumatized. Refugees everywhere are exposed to an increased risk of infection

The same applies to soldiers and mercenaries, for whom wound infections and diarrheal diseases are a common problem. Armed conflicts often increase the risk of HIV/AIDS, for example when mass rape is used as a combat weapon. This has happened in Rwanda, in the Congo, in the war-torn Indonesian province of Aceh, in former Yugoslavia, and in Darfur. In at least some cases in Rwanda the rapists appeared to be aware that they were HIV-positive and blatantly set out to transmit the disease. However, the relationship between HIV/AIDS and armed conflicts cannot be reduced to the long-accepted simple equation "conflicts favor the spread of AIDS." In fact, in some situations the equation "conflicts slow the spread of HIV" is equally true. This assertion is supported by data from Mozambique and Angola. The effect is due to isolation of the population in conflict areas, a reduction of sexual contact in guarded camps, and the close-knit nature of sexual networks of some soldier groups.

In any case, HIV is much more prevalent in many of the world's armies than in the corresponding civil populations. Often soldiers contract the virus through drug abuse and the sharing of needles, as happened in the former Soviet Union during the Afghan War in the 1980s. During the occupation, opium cultivation continued and even increased, and heroin was supplied to Soviet soldiers in order to undermine their morale. Many returned home as addicts, and some had contracted AIDS through the use of contaminated needles.

Here again are two specific examples:

During the genocide in Rwanda in 1994, 800,000 people were brutally slaughtered. Refugees were driven together in camps in Zaire (now the Democratic Republic of the Congo), where cholera broke out. The pathogen was resistant to the standard antibiotics. Within three weeks over 50,000 people died as a result of diarrheal disease. The HIV rate is many times higher among rape victims than among other women. In Rwanda it has been estimated that up to half a million girls and women were raped. As a result, 80% of HIV-positive women now living in Rwanda were rape victims. Similarly, in Uganda, the HIV rates are twice as high among women who were victims of rape.

In Iraq the situation now, five years after the invasion by the USA and its allies, is grim. Official figures report 84,000 dead and 3.5 million refugees, but unofficial estimates are significantly higher at several hundred thousand deaths. As elsewhere, much of the distress in Iraq is due to transmissible diseases, especially diarrheal diseases, wound infections, and tuberculosis. After the immunization programs in the devastated country collapsed, tens of thousands of children contracted measles and mumps. Even among US soldiers in Iraq, more than three-quarters have reported at least one bout of diarrhea. Wound infections also play a role, and multiresistant microbes are a cause of concern. A largely unknown wound microbe, *Acinetobacter baumannii*, has spread in a highly resistant form from Iraq via the transit hospital in Landstuhl, Germany to US hospitals. British soldiers serving in Iraq have also carried the microbe back to their homeland.

The balance teeters

The spread of deadly infectious diseases among the political and military elite of developing countries can undermine the stability of those countries. Foremost among these scourges is HIV/AIDS. As mentioned above, the prevalence of HIV is often significantly higher in armies than among the civil population. As many as half of all soldiers in the Democratic Republic of the Congo may be HIV-positive. Armies have to cope with the disease. Sickness in the ranks means a weaker armed force, and ways of dealing with HIV must be regulated. How are wounded soldiers with HIV to be treated? What economic burdens arise when a large number of soldiers require medication, and families need to be provided for in the event of their death. Should mandatory testing be introduced? These are just some of the questions that need to be answered. HIV can undermine armies in many ways, but in a more subtle way than epidemics that suddenly carry off a large part of the army.

Infectious diseases, especially HIV/AIDS, increase the danger of states becoming unstable. Many scenarios are conceivable. In any case, countries afflicted by infectious diseases find it more difficult to recover from crises. All sectors of the state are affected: education, because children are less likely to attend school and there is a shortage of teachers; healthcare, because the epidemic far exceeds capacities and medical staff themselves fall ill and cannot work. Although the educated social stratum is not affected more severely than the rest of the population, the problem makes itself felt especially rapidly in the thin stratum of the elite, because it already comprises a very small fraction of the overall population.

I've already described the destruction of family structures. The population between fifteen and forty-five years of age are most severely affected by HIV/AIDS. Those left behind are children, who must then make their way through life as orphans, and the elderly, who contribute less to the national economy. There is a theory which claims that adolescent orphans or half-orphans, who often end up as street children, are more easily radicalized. The danger of children ending up doing service in children's armies also increases.

Not least of all in the interests of national security, US President George W. Bush initiated PEPFAR, the President's Emergency Plan for AIDS Relief. Thankfully, there is no evidence of a global security risk in Africa. However, no one knows how long that will remain so or whether a change is already in progress. Rock singer Bono commented: "There are ten potential Afghanistans in Africa, and it is a hundred times cheaper to prevent the fire from flaring up than to stamp it out."

10.4 Hot spot number 3: from the world's laboratories

The attacks on the World Trade Center in New York on September 11, 2001 were followed by several bioterrorist attacks at the East Coast of the USA. Spores of the anthrax bacterium were sent in the form of fine dust in letters. Twenty-two people contracted *Bacillus anthracis* infection, five of whom died. More than 30,000 people were prophylactically treated with antibiotics, and many public buildings had to be decontaminated. These attacks drew attention to a new dimension of the epidemic threat: bioterrorism, the intentional release of pathogens.

The consequences of such an attack would be horrendous. Though different nations assess the actual risk of such attacks and the subsequent outbreak of epidemics differently, there is a common element between the outbreak of epidemics after a bioterror attack and the quasi-natural outbreak of epidemics: Once the pathogen is at large in the world, its spread and transmission follow the same rules in both cases. People play a key role in any outbreak – intentionally in the case of a bioterrorist attack, unintentionally in natural outbreaks – and the spread of the disease also depends largely on human conduct.

Biosafety measures must also include the handling of pathogens in laboratories.

To this end, a code of conduct has been established. The code not only describes procedures for preventing the misuse of pathogens but also sets out general rules for the peaceful handling of hazardous microorganisms. However, a spate of laboratory accidents has demonstrated that safety gaps still exist and that a stronger sense of responsibility needs to be instilled among researchers. Not until July 2007 did an incident come to the public's attention. For the first time the US Centers for Disease Control (CDC) completely withdrew a university's license to handle hazardous organisms. On June 30, 2007, the CDC barred Texas A&M University from working in this field in no uncertain terms. In its letter it stated that it had concerns as to whether the university was observing biosafety standards and whether those responsible were actually familiar with the requirements for handling hazardous organisms and implementing them. What had happened?

While investigating pathogens for potential bioweapons, several employees of the university had become infected with Q fever, a zoonotic disease. One woman developed brucellosis, also an animal disease. Flouting all regulations, the university failed to report these incidents. The CDC only learned of them indirectly. The authorities then ordered the immediate closure of five laboratories with 120 employees. The university was also threatened with the total revocation of its license to conduct research with any infectious agents.

After the SARS outbreak had been brought under control, SARS incidents occurred in more than ten laboratories. Luckily, in no case did the pathogen spread among the population. In another case, a laboratory in the USA was working with insufficiently inactivated anthrax bacteria. It had been erroneously assumed that the bacilli were killed and therefore were no longer infectious. This case also proved to be a lucky escape; no one contracted the disease. In 2005 samples of the H_2N_2-type virus that causes Asian flu was mistakenly sent to several thousand laboratories in eighteen countries – the same viral strain that claimed over one million lives in 1957. Once again, luck prevailed. Not so in the case of a laboratory accident involving the Ebola virus at the Vector Research Institute in Novosibirsk, Russia. A researcher working at the highest safety level accidentally pricked herself with a syringe needle, thus infecting herself with the Ebola virus. After a week she developed hemorrhagic fever; after another week she died. Fortunately, the virus was prevented from spreading to the civil population. The outbreak of foot-and-mouth disease in England in August 2007 is also believed to have been caused by a safety breach at a research laboratory. As worrisome as these accidents are, they were caused unintentionally and ignorantly. In a worst case scenario, a scientist or laboratory worker misuses his objects of research on purpose. As it turns out, this appears to be the cause of the 2001 anthrax attacks. In August 2008 numerous US newspapers including the Los Angeles Times and New York Times reported that the F.B.I. had identified a suspect for these attacks. After they had confronted an anthrax researcher of the biodefence laboratories of the US Army with their accusations, he committed suicide. Were the anthrax attacks due to the yearning of the scientist for increased interest in his field of research? If so, everything panned out as the suspect may have hoped for. In the years following the attacks, almost 50 billion US dollars were committed to biodefence research in the USA and the anthrax vaccine for which the suspect held a patent was further developed with strong financial support from precisely this budget.

10.5 Hot spot number 4: breeding grounds for vectors

The Earth is growing warmer, and the way in which climate change has sparked public debate is unprecedented. Global warming will probably also affect the spread of infectious diseases. For one thing, diseases transmitted by vectors – e.g. ticks, mites, mosquitoes, bloodsucking flies – are likely to spread more rapidly. In addition, higher temperatures favor the spread or diarrheal diseases in general, because the pathogens reproduce in food residues and contaminated water more rapidly at higher temperatures. At the same time, however, other diseases may become less important. For example, shorter, milder winters will shorten the season of coughs, sniffles, sore throats, and flu. Other interactions are also conceivable. Generally speaking, though, we can only surmise such interactions at present. They are too complex to allow us to predict them with any certainty.

Let us examine the scenario in which vectors, which like a warm, agreeable environment, will spread. Malaria will probably benefit most from global warming. It is estimated that a temperature rise of two degrees Celsius will increase the risk of malaria by about 5%; around 100 million more people will then live in areas where malaria is endemic. By the end of the 21st century the proportion of people living in high-risk areas will increase from 40 to 60%. At the same time the global population will have grown tremendously. The disease will mainly spread upward, not outward. Thus mosquitoes will also be able to thrive in the highlands of Kenya and the higher regions around Mount Kilimanjaro. Moreover, the malaria season will be longer in many countries in which malaria is already endemic. The pathogen itself takes about 26 days to mature at 20 °C but only about half as long at higher temperatures. The optimum water temperature for the mosquitoes to reproduce is around 28 to 32 °C. The warmer it becomes, the more often the mosquitoes go in search of a blood meal. More pathogens, more vectors, and more frequent feeding by the mosquitoes all spell an increased risk of infection.

Other insect-borne tropical diseases, such as Chagas disease, sleeping sickness, various forms of leishmaniasis, river blindness, and elephantiasis, could also spread as a result of rising temperatures. Again, mainly higher lying areas would be affected. However, evidence of the influence of global warming on these diseases is rather vague. In the case of schistosomiasis, which is transmitted via snails, there are contradictory scenarios. Rising temperatures could cause the disease to spread in some regions, but the opposite is also possible if the water becomes too warm for the snails to thrive.

Foremost among the vector-borne diseases that will benefit from global warming is dengue fever, the most important mosquito-transmitted viral disease in the world. The mosquitoes have already adapted to life in big cities. They thrive best in puddles, for example in old tires or tin cans. Thus, if the environment becomes not only warmer but also wetter, the area infested by the dengue mosquito could spread.

An instructive example is the occurrence of West Nile virus in New York. The virus is transmitted by mosquitoes from birds to humans. And these mosquitoes thrive especially well in standing bodies of water near cities, for example in parks and on golf courses. In 1998 and 1999 the winter in New York was particularly mild, and the mosquitoes were able to breed in large numbers. In the following dry summer numerous birds gathered at the remaining bodies of water, where they were bitten by the mosquitoes, so that the microorganisms spread among the bird population. Consequently, New York saw an outbreak of encephalitis and meningitis. Six elderly people died. Meanwhile the West Nile virus has spread throughout the USA. More than 20,000 cases have so far been reported. Speculation is rife about how the virus was originally introduced. Explanations range from bioterrorist sabotage to infected travelers from the Near East to imported exotic birds.

In Europe the West Nile virus reservoir has been essentially limited to birds and has only sporadically jumped to humans. Probably mosquitoes in Europe bite either humans or birds, whereas those in the USA see both as sources of blood meals. An interesting example is the spread of Chikungunya virus in the

Chikungunya fever – disease outbreak in vacation paradise

In 2005/2006 a largely unknown disease broke out on several islands in the Indian Ocean, including La Réunion. It was soon identified as Chikungunya fever. The disease is caused by RNA viruses known as alphaviruses and is transmitted by mosquitoes. Chikungunya means "crooked," which aptly describes the clinical picture. Victims develop severe joint pain, which may be accompanied by severe complications such as liver damage or meningoencephalitis. Asymptomatic cases are rarely observed, i.e. infection virtually always manifests itself in symptoms. Chikungunya fever was first described in Tanzania and Uganda in the 1950s. By the end of the 1950s it had spread as far as Southeast Asia. The new epidemic broke out in Kenya for the first time in 2004 and spread to the Comors in 2005, where travelers or the mosquitoes carried the virus to the islands in the Indian Ocean. The islanders on La Réunion probably had no immunity whatsoever against the disease, because it spread through the population like wildfire. More than a third of the 770,000 inhabitants contracted the disease in 2006, with 266,000 clinical cases. Two hundred and fifty people died. Afterward the disease reached India, where 1.3 million cases have already been reported. It is estimated that around two million people contracted Chikungunya fever in 2006.

Some time in 2005 a new virus strain arose in Kenya or on the Comors. Evidently, this made it easier for the virus to colonize another vector, the Asian tiger mosquito. This mosquito, which also transmits dengue and yellow fever, was already present on La Réunion. Thus, Chikungunya fever can potentially spread to other continents. As mentioned, it has already reached India, and in September 2007 the first cases in Italy were reported.

Indian Ocean, which is described in the box entitled "Chikungunya fever: disease outbreak in vacation paradise." In our latitudes the warmer weather is especially beneficial to ticks, the result being that the risk of tick-borne diseases is on the increase.

But not only insects thrive in warm environments. Rodents such as rats and mice also love warmth. Rodents carry numerous infectious diseases – from hantaviruses to bubonic plague – most of which are transmitted to humans via fleas.

Some of the microorganisms themselves will benefit from rising temperatures. Chief among these, in addition to agents that cause diarrhea, are meningococci. The fact that long droughts favor the spread of meningococci is well established. The warming of the Earth's seas and the increased frequency of storms and floods could help spread cholera further, as the bacteria can hitch a ride on algae. Warmer water could promote algal growth and thus spread the disease.

To sum things up, global warming may promote infectious disease through a range of mechanisms, many of which involve vectors as intermediaries. However, as far as we can determine today, there is little cause for concern that global warming will cause new epidemics to break out unpredictably.

10.6 Hot spot number 5: face to face with the wilderness

Every year the Earth loses about 13 billion hectares of woods to clearing, mostly in the tropics, though recently some areas have been increasingly reforested. The largest forest areas lost, 4.3 million hectares a year, are in Latin America, followed by Africa. Often the purpose of clearance is to make the land suitable for agricultural purposes. Usually canals and dams, in which insects thrive, are also built for irrigation. For this reason, defor-estation in tropical areas is often a prelude to vector-transmitted infectious diseases such as malaria. Paradoxically, reforestation can also provide good breeding grounds for infectious diseases. Thus, reforestation in New England in the USA has facilitated the re-establishment of red deer. Along with the deer, borreliosis-transmitting ticks also proliferated, leading to an outbreak of Lyme borreliosis.

Humans come into especially close contact with the wilder-ness when felling individual trees, usually species that provide high-grade wood. First, the workers come into close contact with wild animals. Second, they hunt wild game as a source of food. Third, they keep captured wild animals as domestic pets. HIV, Marburg virus, and Ebola virus are prominent examples of microorganisms that have infected humans in this way.

The roads and tracks built for transporting logs also make it easier for poachers to hunt. Loggers are often also part-time poachers or become full-time poachers after they have lost or given up their job as loggers. In southern Cameroon, for example, three out of four poachers are former loggers. The poachers sell the animals they kill as bushmeat. Far from being a rarity, in some regions such meat is a staple in the diets of the indigenous people. In the Congo Basin, for example, the inhabitants eat on average 280 grams of bushmeat, amounting to a whopping 4.5 million tons annually. Poachers not only decimate already threat-ened species, such as bonobos and gorillas (see box: "Bushmeat: from the tropical forest to the table"), hunting our close relatives is also fraught with danger. Because of their close kinship with us, the species barrier which a pathogen has to cross in order to colonize humans is potentially low.

Ebola virus and HIV are examples of how both acute and chronic infections can be transmitted. Because of globalization,

Bushmeat: from the tropical forest to the table

Cannibalism is a relatively rare phenomenon among humans. However, our closest relatives do often end up in the soup pot or on the grill, especially in Africa. Monkey meat, along with the muscles and entrails of other wild animals, is sold as bush meat in many markets. Today the entire African continent is home to around 200,000 chimpanzees, 50,000 bonobos, and 120,000 gorillas. In northern Congo alone it is estimated that 5 to 7% of the indigenous chimpanzees and gorillas are slaughtered every year. Because the animals, which are often under protection, breed slowly, the losses are irreplaceable. If the consumption of ape meat continues at its current pace, our closest relatives will face extinction. At the same time, humans are creating a serious health risk for themselves: apes carry numerous pathogens from the remotest regions. And because the animals are so similar to humans, it is often easy for viruses and bacteria to jump across the species barrier and infect humans. The list of potential diseases is long. It includes such dread diseases as Ebola hemorrhagic fever, whooping cough, anthrax, encephalitis, smallpox, hepatitis, influenza, measles, mumps, and rubella. Three quarters of all new infectious diseases are zoonoses, i.e. they originate in animals. HIV has driven home the fact that the threat is not merely a fanciful scene from a B horror film. In all probability the deadly virus was transmitted to humans from monkeys. HIV-2 probably came from viruses that were transmitted from sooty mangabeys directly to humans. HIV-1 was probably transmitted by chimpanzees. However, immunodeficiency viruses are harmless for many ape and monkey species. More than 20 simian immunodeficiency viruses have so far been found in apes and monkeys in Africa. It is possible that the precursors of HIV did not initially cause symptoms in humans either but eventually transformed into the deadly AIDS virus through a series of mutations.

Because HIV infection does not immediately lead to the full-blown disease, the virus was able to spread around the globe undetected. It then unleashed an epidemic of the worst kind imaginable.

Ebola outbreaks following contact with apes have also captured headlines around the world on several occasions. In the 1990s the deadly virus infected humans on three occasions in the central African country of Gabon. In these three outbreaks 142 people contracted the disease, of whom two-thirds died. Unlike HIV, however, Ebola rapidly manifests itself as a life-threatening disease. We might therefore conclude that it is unlikely to spread to other countries and that the problem will remain local. So far this has been the case. On one occasion an infected doctor flew from Gabon to South Africa and transmitted the virus to at least one other individual, who, like the doctor, also died soon afterward, so that a large-scale outbreak was nipped in the bud.

But it could have ended quite differently. Monkeys are regarded as delicacies, even by many Africans living in Europe and America. And there is a thriving black market for monkey meat in many big cities in Europe and North America. Illegal markets can be found in New York, Chicago, Toronto, Montreal, London, Brussels, and Paris. Every month around 6000 kilograms of illegally killed game is sold on these markets. Fifteen percent of the produce is monkey meat, and 1% comes from chimpanzees, bonobos, and gorillas. What would happen if a delivery were contaminated with Ebola virus, which then infected a cook or kitchen helper in a metropolis like New York? Richard Preston vividly describes what could happen in his thriller *The Hot Zone*.

even the deepest jungle areas are now potential sources of pandemics. Green monkeys legally exported from Uganda were responsible for the Marburg virus outbreak in the Marburg Behring plants in 1969. Infections were later also reported in Frankfurt and Belgrade. A total of 31 people became infected, of whom seven died. At the end of the 1990s the Marburg virus broke out in the Democratic Republic of the Congo, claiming over 100 lives. At the turn of the year 2004/2005 Ebola flared up in Angola, claiming more than 300 lives. In July 2008 a Dutch tourist soon after her return to The Netherlands from a trip to Uganda died of Marburg virus infection – the first case in Europe for decades.

The exportation of exotic animals has taken on enormous proportions. The market for exotic animals as pets is flourishing, and not always legally. A motley assortment of exotic animals arrive at the major airports of Heathrow in London, J. F. Kennedy in New York, and Schiphol in Amsterdam, where they remain awaiting further transport. Inter-animal contacts are possible at these centers that would never occur in nature. Within a small space an exotic bird from Asia meets monkeys from Africa and reptiles from South America. Unparalleled opportunities arise for billions of microbes to "learn" to overcome barriers in order to infect new species.

Around 200 million fish, 50 million amphibians, two million reptiles, 350,000 birds, and 38,000 mammals were introduced into the USA alone in 2003. In 2003, seventy-nine people suddenly contracted monkey pox, which they had contracted from prairie dogs they kept as pets. The viruses had originally entered the country in Gambian giant pouched rats. They then infected prairie dogs and were transmitted from the prairie dogs to their owners.

10.7 Hot spot number 6: mankind and all creatures great and small

Planet Earth is home to 6.5 billion people. In addition, there are
1 billion pigs; 3.3 billion cattle, buffalos, sheep, and goats; and 30
billion poultry. All these animals are bred for our consumption. In
addition, every year nearly 200 million tons of fish, crustaceans,
snails, mussels, and other shellfish are harvested. We are aware
that we are eating too much meat, but it has not disappeared
from our diets. Although meat consumption has dipped slightly
in the industrialized countries, this is more than made up for by
a marked boost in developing countries. Greater meat consumption is a sign of prosperity. In 2006 global meat consumption rose
by 2.5%. China has emerged as the world's biggest meat producer. In 1980 it produced one-tenth of the world's meat; by 2004
that figure had climbed to just under one-third. A similar trend is
occurring in Brazil, which accounted for less than 4% of global
meat production in 1980 and nearly twice that amount in 2005.
At the same time, consumption has declined in the industrialized countries. Germany, France, and the Netherlands together
accounted for a good 10% of global meat production in 1980 but
just under 6% in 2004.

The world's enormous meat consumption is made possible
largely through animal farming on a massive industrial scale.
Fourteen billion poultry are languishing in China, nine billion
in the USA. The Netherlands, a major producer of animal feed
in the EU, is home to 16 million people along with 11 million
pigs and 90 million poultry. One of the fastest growing markets
worldwide is aquaculture, the mass cultivation of fish and shellfish. In 2005 just under 60% of fish catches were straight from
the sea. By 2030 well over half of the fish consumed will come
from aquaculture farms. That is already the case for some fish
products: three-quarters of salmon and nine-tenths of mussels on

the market were bred on aquaculture farms. Fish can be penned together very tightly; after all, aquiculture allows crowding in three dimensions. The higher risk of infections in aquacultures is then countered by the heavy use of antibiotics in feed.

Why is this a cause for concern? The problem has already been outlined in Chapters 5 and 6. On industrialized farms animals are packed tightly together. They are bred for maximum meat production, not for resistance to pathogens. Hygiene in many operations is bad to catastrophic, and pathogens are able to spread, literally, like the pest. H5N1 outbreaks on poultry farms are just one example. To keep animals' health under control, antibiotics are added to the feed in many countries. The same applies to aquaculture. In addition, antibiotics are used to promote growth. In the USA 70% of all antibiotics are used in animal farming. Of all antibiotics produced annually, half are used to promote growth and, to a lesser extent, to prevent illness in farmed animals. It has been estimated that 25 million pounds of antibiotic is produced annually for livestock treatment. As a result of this practice, resistant strains of microorganisms are rapidly developing. This relates primarily to bacterial infections, but some resistant viruses have also come from animal-breeding farms. Thus the addition of amantadine and rimantadine to the feed of Chinese poultry batteries has resulted in the emergence of influenza viruses resistant to these drugs. Yet the costs of doing away with antibiotics in animal farming would not be an unbearable burden on consumers. In the USA a ban would translate into extra costs of not more than two cents more per pound for chicken and six cents more per pound for pork or beef – about ten dollars a year.

In addition, animal farming creates staggering amounts of excrement. A poultry farm housing several tens of thousands of chickens quickly produces a ton of excrement. The same applies to pig farming. A single pig produces up to two tons of slurry a year. If the stock becomes infected and the waste is not properly disposed of, disease can spread rapidly.

Europe has introduced strict regulations on animal farming and the use of antibiotics. In the USA the laws are somewhat less restrictive, and in many developing and threshold countries they are nonexistent. Meanwhile, the meat market has taken on global proportions. Meat is moved back and forth across national borders. A vigorous trade has also sprung up between Europe, North America, Latin America, Africa, and Asia. Industrially farmed chickens and laying hens have become the migratory birds with the greatest range. China has emerged as the major poultry supplier for West Africa, exporting one million chickens a year to the region. Unfortunately it is not all done legally. An H5N1 outbreak in Nigeria in 2006 was very likely caused by smugglers who brought young chicks into the country to be fattened into roasting chickens.

Animal farms are also a breeding ground for new pathogens that could jump the gap to humans. This scenario is already causing concern to experts, especially in the light of several recent incidents. Consider SARS, bird flu, and BSE, to take just three examples. The transmission of BSE to humans produces variant Creutzfeldt-Jakob disease. BSE has so far been reported in more than 30 countries, and tens of thousands of cattle have been slaughtered. Over 190 people in the EU have developed variant Creutzfeldt-Jakob disease.

The most consumers can do is make sure that they cook or boil their meat and eggs. Many microorganisms are killed in this way. However, animal farms might not necessarily be the direct source of a pathogen that unleashes a pandemic. It could arise not only among workers on animal farms but also in slaughterhouses and in the meat processing industry, where animals from a large number of farms converge. Humans come into close contact with organs during the slaughtering process. One sick animal is enough to contaminate whole shipments of meat. Finally, in the last link of the chain, waste is not properly disposed of in some countries. Blood and intestinal matter come into direct contact with bodies of water, creating new sources of infection.

10.8 The next pandemic

Where do I see the gravest threat of a new pandemic? The growing contact between humans and animals – both in the wild and in industrialized farming – alarmingly increases the risk of new pathogens, one of which could have the makings of a wholesale killer. Even if this occurs distantly in Asia, Africa, or somewhere in the back of beyond, we cannot feel safe. We live in a global village with 800 million tourists, 1.5 billion flying passengers, and millions of legally and illegally exported animals, every year.

Of the approximately 150 new diseases that have emerged in recent decades, around three quarters of them come from animals, i.e. they are zoonoses. In over 40% of cases the causative agents are viruses: 30% are caused by bacteria, 10% by protozoa, and 5% by worms and fungi. Viruses appear especially well poised to overcome the species barrier. In all probability the agent of the next pandemic will be an RNA virus. RNA viruses first have to translate their information into DNA, and, because it is far from perfect, this process affords enormous scope for mutations. Recombination exchanges between related strains can further exacerbate the risk.

Perhaps the new infectious agent will carry a receptor that allows it to dock to a conserved structure on human cells. In this way the virus could jump to any foreign species having the same docking site – an ideal setting for it to adopt a variety of hosts. Such a pathogen is most likely to occur in a hot spot where humans and animals "rub shoulders" and in a region undergoing an ecologic, demographic, and social transformation. Places where I could easily imagine it happening are mass animal farms in Asian threshold countries and the deforestation areas in central Africa. The world's mobile population will then carry the pathogen around the globe.

Not every attempt of a microorganism to jump from animals to humans is successful. On the contrary, most attempts result in the microorganism's demise. But they never give up. Eventually a mutant is sure to establish a bridgehead in humans. We believe that HIV-1 and HIV-2 attempted to make the jump to humans at least ten times before meeting with success. At the moment Spuma viruses and Nipah viruses have launched an especially determined bid to make the jump from monkeys to humans – so far with little success. But it is probably just a matter of time. A horror scenario would certainly be a pathogen, like HIV, that first insinuates itself latently in humans, to then spread from person to person in the air like influenza viruses and tubercle bacilli. SARS has already made a foray along these lines.

Many of our thought processes and actions follow linear patterns. Epidemics, however, grow exponentially in every phase. A bacterium replicates by dividing at a rate of about one division every 30 minutes. If a bacterium has spread to cover half a culture dish in three days, it does not take another three days for it to colonize the entire dish, as you might think, but just 30 minutes. A cell infected with a virus easily produces 10,000 viral offspring, each of which is capable of infecting a new cell. It is the same with epidemics. They may start slowly and tenuously, but once a microorganism has taken hold in humans and learned how to infect other people and once a critical number of individuals have been infected, it then progresses with breathtaking speed. It is a bit like snowflakes building up into avalanches.

10.9 Global threats call for global responses

The world has been shaken awake by fears about SARS and H5N1. With regard to the early detection and immediate prevention of infectious diseases, the prevailing view is that global threats call for global responses. June 15, 2007 marked a milestone in global healthcare policy. On that day the member states of the WHO widened the health regulations that had been in place since 1969. The overriding aim of the move is to prevent a national health crisis from becoming global. The aim of the regulations is to prevent local endemics from spiraling into global pandemics. Besides cholera, yellow fever, and bubonic plague that were already covered, the list now also includes smallpox, polio, SARS, and new influenza strains. They also include all other potential outbreaks as well as biological, chemical, and radioactive threats in order to be as flexible as possible.

The signatory states undertake to report potential outbreaks without delay and to recognize the coordinating role of the WHO in countering global threats. National and international monitoring and surveillance systems for the early detection of outbreaks are to be set up and expanded. Safety precautions are standardized at international ports of entry, i.e. at airports and shipping ports. A coordinating national center, standard guidelines for assessing threats, and the evaluation of the required countermeasures also facilitate rapid response. Protection of life and health takes precedence over unrestricted trade.

At the time of writing, Indonesia is refusing to supply the

WHO with H5N1 virus samples. Such samples would provide insights into the development of the pathogen and its resistance to drugs. They are not only needed as indicators for an early warning system but would also facilitate the rapid development of an effective vaccine should bird flu make the final jump to humans and embark on widespread human-to-human transmission. Indonesia fears that it will be deprived of the benefits of developments derived from the strains, and that the vaccine will be patented and made unaffordable for developing countries. Under the new health regulations Indonesia's refusal to provide virus samples is illegal. Yet, Indonesia's fears are not so far-fetched: fewer than a dozen industrialized countries have production facilities for influenza vaccines. Because of limited capacities, it is feared that, in the event of a pandemic, the industrialized countries may ensure timely supplies to themselves, while developing and threshold countries are left empty-handed.

Should a real threat arise, it is hoped that the international health regulations will be rapidly implemented and will not take a back seat to self-serving trade interests. As always, the signatories have been granted a period in which to translate the regulations into national law: until 2012 or, in exceptional cases, until 2016.

11 Five To or Five Past Twelve?

The question today is how we can persuade mankind to agree to its own survival.

Bertrand Russell (1872–1970), English philosopher and mathematician,
Nobel laureate for literature 1950

If you think you're too small to make a difference, try sleeping in a closed room with a mosquito.

From Africa

11.1 Controversial but convincing

The city of Siem Reap in Cambodia is only a short distance from the impressive temples of the early kingdom of the Khmer. The Angkor archeological park, with Angkor Wat, Bayon, and many other temples, has been awarded World Heritage status by UNESCO. However, I had a different reason for going to Siem Reap: I went there to visit the Jayavarman Children's Hospital, run by the Swiss pediatrician Dr. Beat Richner, who also arranges the financial support for the hospital. Through his Kantha Bopha foundation he channels private donations to this and three other children's clinics in Phnom Penh, Cambodia's capital, as well as to an obstetric clinic for HIV-positive mothers. Ninety percent of the total costs are provided by the foundation while the Cambodian state puts up the remaining 10%. In 2006 the Kantha Bopha clinics hospitalized and treated 96,000 children with serious illnesses, provided one million people with out-patient treatment and administered 320,000 basic vaccinations. Eighty-five percent of all Cambodian children receive help from the four clinics. Beat Richner wrote me that in early 2007 the country was once again plagued by dengue fever. By the middle of the year the clinics had already taken in and treated 14,000 children with, or on the verge of, dengue shock syndrome. Dengue is not the only infectious disease that ravages Cambodia. Two hundred thousand people are HIV-infected, 10,000 of them children. Six times that number have already lost one or both parents to AIDS. Every year, 70,000 Cambodians contract tuberculosis. A large number of children are forced into prostitution. Many of them take yama, a methamphetamine drug. Not infrequently they fall victim to pedophile tourists.

Abject poverty is rife in Cambodia. Between the 1970s and 1990s, for more than a quarter of a century, the country was afflicted by civil wars, military occupation and genocide. Millions of people lost their lives under the harsh regime of the Red Khmer. Only in the early 1990s did the situation calm down, and in the past few years the economy has begun to show tentative signs of growth. Siem Reap is now benefiting from tourism. But even here, the visitor still cannot help noticing the large number of physically disabled people – some of the 40,000 Cambodians who have been mutilated by land mines. Even today twenty to thirty people a month are still seriously injured by mines.

Inside the Jayavarman children's clinic the contrast to the outside world could hardly be more stark. The cleanliness is unimaginable. There are numerous openings directly under the roof which ensure that the air is continuously refreshed. Hope prevails here, although you can see and sense the suffering of the sick children. The sick are examined and treated with the latest medical technology that we would expect to find in every rural clinic but is still largely unknown in the developing world. This modern care is provided free of charge. Parents can also stay with their children, take care of them, and comfort them for the entire time they are hospitalized – again, without having to pay. The vast majority of Cambodian families could hardly afford even one day in hospital. By not charging for hospitalization or treatment, Beat Richner has pitted himself against the opinion of the major aid organizations, including the WHO. These believe that patients should pay for their health themselves and that medical care should reflect a country's economic situation. Beat Richner considers it discriminatory to deprive the poor, and especially children, of the best possible preventive measures, diagnostics, and treatment available. He believes that they too should have unrestricted entitlement to the universal right to health. To ensure that this is so, the staff at the clinics are paid much higher wages than what is normal in Cambodia. This way they do not have to resort to accepting bribes, and general morale is very good.

Financed with private money, the physicians and nurses at the clinics treat 25,000 children newly infected with tuberculosis every year. Many cases can only be diagnosed with the help of modern medical technology. For the patients this would be totally unaffordable, as would the months of treatment required to cure the disease. The foundation also assumes travel expenses and the cost of monthly checkups – many patients come from far away. Ninety percent of the patients continue with the treatment until it is finished. The HIV cases also prove Richner right. In the clinics HIV-positive mothers are given ART, and the babies are born by caesarian section and are fed ready-to-use milk products instead of mother's milk. The upshot is that for the newly born the diagnosis is almost always "HIV-negative."

Even if Beat Richner is controversial, his combination of Swiss clockwork precision, humanitarian thinking, and readiness to fight for his ideas whenever he has to deserves our admiration. I asked him how he managed to cope with it all – getting hold of funds, lending a hand in the clinic, organizing the clinics at the same time, and constantly having to experience so much depressing suffering. I received an answer that weekend: to assuage his bitterness he plays cello Saturday evenings for the tourists in Siem Reap. And it even brings in donations.

11.2   Expensive but still affordable

Figures are nearly always abstract. Who can really grasp the fact that 15 million people die of infectious diseases every year? Even 15,000 is a difficult quantity to imagine. What is even harder to comprehend is the immense loss of life-years due to sickness, disability, and early death. In the end DALY (disability adjusted life years) is an artificial term, although a very useful one (Fig. 23). But to conclude, let's talk about figures again – partly because, aside from the humanitarian dimension, people always immediately cite the costs. In fact, billions of euros will have to be invested worldwide to bring epidemics under control. What is needed are epidemiological, technological, and medical measures to identify and control new epidemics in their early stages and to suppress existing epidemics. In addition, we need to step up research and development for new intervention measures against those infectious diseases for which we currently have insufficient preventive and therapeutic options. And last, but not least, we must tackle the causes of epidemics and combat both general causes, such as poverty, and specific ones. The problems are global, and the global economy counts in billions as a matter of course.

DALYs of the most important infectious diseases (years lost due to disease)

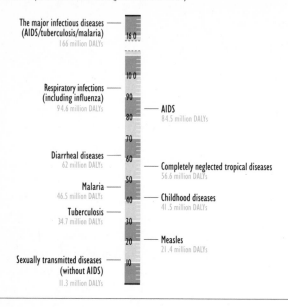

The major infectious diseases ——
(AIDS/tuberculosis/malaria)
166 million DALYs
                                                           160

                                                           100
Respiratory infections ——
(including influenza)   90
94.6 million DALYs   80   —— AIDS
                                                                               84.5 million DALYs

                                                           70
Diarrheal diseases ——
62 million DALYs   60
                                                                           —— Completely neglected tropical diseases
                                                                               56.6 million DALYs
                                                           50
Malaria ——   40   —— Childhood diseases
46.5 million DALYs   41.5 million DALYs
Tuberculosis ——
34.7 million DALYs   30

                                                           20   —— Measles
                                                                               21.4 million DALYs
Sexually transmitted diseases ——   10
(without AIDS)
11.3 million DALYs

Figure 23 DALYs of the most important infectious diseases

It won't come cheap. But perhaps we should first put the costs in perspective. In 2006 the world invested 900 billion euros in armaments, 137 euros for every man, woman, and child on the planet. Spending on perfumes and cosmetics is roughly as high as the annual amount for the Millennium Development Goal in the field of health and health care (Fig. 24). SARS alone cost three times as much as the annual budget of the Global Fund to Fight AIDS, Tuberculosis and Malaria. Every year in the USA influenza alone gives rise to total costs that exceed the budget of the Gates Foundation by one-third. To develop new drugs and a vaccine against tuberculosis we would only have to invest as much money over a ten-year period as adult entertainment in the USA gobbles up in one year. US citizens spend two billion dollars a year on keeping their teeth white. With this money 500 million persons suffering from a neglected tropical disease could be treated over five years. According to the economist and Nobel laureate Joseph Stiglitz, US citizens have to pay 13.5 billion dollars every month for the US presence in the Iraq. I am quite aware that this sort of comparison can easily be written off as "populist." But the list really is impressive – and I could go on adding to it. There can be no doubt that to control infectious diseases, the richer countries are going to have to pick up more of the tab than their poorer counterparts. The return on this investment will be some time coming. But the near eradication or effective control of measles, mumps, rubella, polio, diphtheria, and tetanus as well as the success of the newer generation of vaccines against hepatitis B, *Haemophilus influenzae* type b, meningococci, and pneumococci are meanwhile yielding five to twenty euros for every euro the measures have cost. This is because no costs are incurred for hospitalization or outpatient treatment or diagnostic and therapeutic agents. Even so, there is still an imbalance in the savings – but this time the shoe is on the other foot. For the cost savings are achieved not so much in our latitudes as in the countries once hardest hit. But that is certainly no reason for doing nothing.

As a global problem, infectious diseases have to be tackled globally. For the first level we have to "think global, act local." However, it is just as important to "think global, act global." In the following I would like to share my thoughts on boosting the fight against infectious diseases with you. First I will address proposals for change at a global level. I will then point to a few success stories, for I believe they can serve as examples for action that could be taken elsewhere. Finally, I will present virtual

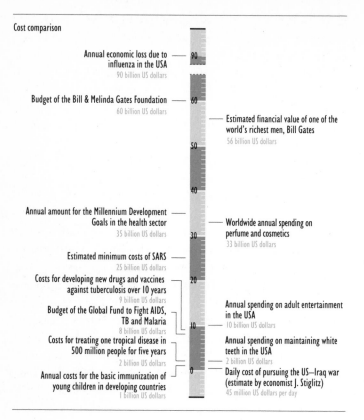

Cost comparison

Annual economic loss due to — 90
influenza in the USA
90 billion US dollars

Budget of the Bill & Melinda Gates Foundation — 60
60 billion US dollars

— Estimated financial value of one of the
world's richest men, Bill Gates
56 billion US dollars

50

40

Annual amount for the Millennium Development —
Goals in the health sector   30
35 billion US dollars

— Worldwide annual spending on
perfume and cosmetics
33 billion US dollars

Estimated minimum costs of SARS —
25 billion US dollars

Costs for developing new drugs and vaccines — 20
against tuberculosis over 10 years
9 billion US dollars

Annual spending on adult entertainment
in the USA
10 billion US dollars

Budget of the Global Fund to Fight AIDS,
TB and Malaria   10
8 billion US dollars

Costs for treating one tropical disease in
500 million people for five years
2 billion US dollars

Annual spending on maintaining white
teeth in the USA
2 billion US dollars

Annual costs for the basic immunization of — 0
young children in developing countries
1 billion US dollars

Daily cost of pursuing the US—Iraq war
(estimate by economist J. Stiglitz)
45 million US dollars per day

Figure 24 Comparison of costs due to infectious diseases with other
expenditures

scenarios that exemplify our options and their consequences.
None of this claims to show the full picture; it is merely intended
to provide food for thought. We need not only an unconditional
will to take action but also imagination and a goodly helping of
creativity. Nature shows us the way. Who, 50 years ago, would
have had the imagination to believe that pure proteins could turn

into pathogens and transmit disease, i.e. multiply and infect? In 1982 Stanley Prusiner set up his prion hypothesis to which the scientific community reacted with intense skepticism. Today it has been established that BSE and the variant Creutzfeld-Jakob disease are triggered by rogue protein particles. Health is a global public good and is well worth making an effort for.

11.3 Everyman's right, everyman's duty

It's a noble concept: global public goods that are available to all and denied to none. They are not the object of rivalry and can be used by many people at the same time. The Earth's atmosphere is one such, as are peace, security, justice, a sound environment, culture, and functioning business and financial markets. And also health and protection against disease. The trouble is that no one feels directly responsible for preserving something that everyone can use but no one really owns. In the final analysis, in terms of transnational epidemics and their control, this means that national containment strategies are only effective, at best, in the short term. Effective control only works on a global scale and best of all if the threat of epidemic is immediately identified locally where it first arises and comprehensive measures are implemented with worldwide support. Hence, the global good of protection against infectious disease not only implies everyone's right to protection, but also everyone's obligation to do what they can. The good news is that occasionally this has already worked. Smallpox has been eradicated thanks to global efforts, while polio has been forced down to fewer than 2000 cases and measles to fewer than half a million.

11.4 Act globally

I will restrict myself to a few examples and hope that these and other innovative initiatives for global action are implemented. Some have already been mentioned, such as an amendment of patent law and the unequivocal clarification of the TRIPS Agreement (cf. Chapter 9). My examples are listed in the box entitled "Ten-point program to control infectious diseases":

Vaccine bonds

I have already mentioned GAVI, the Global Alliance for Vaccines and Immunization, on a number of occasions. This PPP provides funding for vaccination campaigns in more than seventy of the world's poorest countries with a cleverly devised financing concept. To support the GAVI programs, six industrialized countries – Great Britain, France, Italy, Norway, Spain, and Sweden – have launched a finance facility with a budget of more than four billion dollars over ten years. One billion dollars have been raised since 2006 with 854 million dollars disbursed by early 2008. This money is to support the immunization of more than 500 million children and save ten million lives. The novelty is that GAVI can already use the money before the governments have paid, i.e. draw on development aid funds in advance. To do so, international capital markets are tapped, i.e. bonds are issued against legally-binding commitments on the part of sovereign donors to provide aid funds. The commitments of the donors serve as collateral. The whole enterprise is organized by what is known as an International Finance Facility. This passes the proceeds from the bonds to GAVI, which in turn uses them to fund approved immunization campaigns in recipient countries by means of nonrepayable grants. The bond creditors therefore provide interim financing. Private investors can participate just

Ten-point program to control infectious diseases

- Make intensive use of available intervention measures – besides vaccination and chemotherapy also nonmedical technologies – to push infectious disease in the developing world down to their levels in industrialized countries.
- Combat poverty and set up a well-functioning healthcare system coupled with debt relief.
- Set up state-backed funds for nongovernment investors to finance immediate support of infectious disease control.
- Levy international charges to finance infectious disease control.
- Support targeted research into, and development of, drugs, vaccines and nonmedical technologies in order to redress the 90:10 imbalance.
- Impose affordable prices for vital drugs in all countries on Earth and support this with amendments to patent laws as well as through an improved TRIPS Agreement.
- Strengthen PPPs by encouraging governmental and nongovernmental organizations, the pharmaceutical industry, foundations, and other bodies to join forces.
- Set up an international monitoring and surveillance network under the central coordination of the WHO for the early identification and control of new outbreaks of infectious disease.
- Foresight is better than hindsight: the threat of a new outbreak of infectious disease due to man's meddling with nature is considerable and can only be tackled by a change of heart regarding industrialized livestock production and how we manage the wilderness.
- Harness the power of imagination, innovation, unorthodox thinking, the courage of our own convictions.

the same as banks, fund managers, pension funds, insurance companies, and the like. The bonds have been triple-A rated and are therefore rock solid. Ultimately, part of the reason for this is that only about 70% of the committed funds are used as collateral, the rest serving as a cushion against risk. The sovereign donors pay the interest and redemption on their own, taking up long-term debt in international financial markets. Critics of this model have coined the term "development aid on credit" and criticize, among other things, the administrative costs. Be that as it may, it is an innovative approach that makes money available quickly and expendable.

Taxes for the major epidemics

The discussion about tax-financed development aid is not new. For example, France, Brazil, Norway, Great Britain, and Chile plan to levy a tax on air tickets and make the revenues available to combat AIDS, malaria, and tuberculosis. As the tax is charged to all passengers who board a plane in these countries, it affects all airlines in equal measure. When the program is in full swing, each passenger in tourist class will be charged an additional six dollars per international flight, while business-class travelers will have to cough up an extra 25 dollars. Up to 12 billion dollars could be collected this way. To make sure that the money is invested effectively, the Clinton Foundation (launched by former US President Bill Clinton) is negotiating with pharmaceutical companies to secure affordable prices for pharmaceutical products. So far the project has not really gotten off the ground: the expected income for the first year comes to 300 million dollars. At first Germany supported the initiative, but later reneged as the country's parliament refused to accept the tax. Similarly, there have often been like proposals for a tax on arms spending. There are numerous alternative ideas.

Swap debt for health provision

At the beginning of this century the sixty poorest countries had run up debt to the tune of about 540 billion dollars, up from 25 billion in 1970. Over thirty years the poorest countries had paid off 550 billion dollars of debt – and borrowed a great deal of fresh money. Meanwhile the developing countries are paying thirteen dollars on debt service for every dollar they receive from creditor countries.

Debt relief is either not taking place or is moving far too slowly, and poor countries will never be able to shake off the burden. No wonder that debt relief is regularly on the agenda at G8 summits. As early as 2005 at Gleneagles the G7 countries promised to write off 14 billion dollars of debt. The few countries that actually received relief, such as Zambia and Tanzania, have invested part of the money they gained in improving their healthcare systems. So why not turn this into a principle and instead of simply providing debt relief, swap debt for health provision?

It could look like this: in the private sector, debt that is more or less written off is marked to market and transferred to specialized investors, who then negotiate the reduced value with the debtor. Let's assume that a given country has 25 million dollars of debt that it is unlikely ever to be able to pay off. Why shouldn't the former creditors (an industrialized country, the World Bank, or the World Trade Organization) hand over a portion of the sum to an aid organization? Let's further assume that in this way five million dollars is invested in health. The rest of the debt, 20 million dollars, is written off. A key motivation for the debtor countries is that their obligation toward a foreign creditor is transformed into an undertaking toward their own people. Needless to say, swapping debt for spending on healthcare also presupposes considerable responsibility. Transparency must be guaranteed, corruption ruled out. These are by no means new ideas; they have already been practiced occasionally but then discarded. I regard them at any rate as an important weapon in the battle against poverty and disease and a key to the strengthening of the economies of developing countries. I am therefore all the more thrilled to learn that Germany, together with the Bill and Milinda Gates Foundation and others will pursue this very model in the form of the Debt2Health debt-relief program. A start will be made by Indonesia, which Germany will excuse 50 million euros of debt on condition that it invest 25 million euros in its health system. Peru, Pakistan and Nigeria are to follow. Similarly, the UK has promised to contribute 700 to 960 million dollars towards debt relief over the next ten years.

11.5 It can work

There certainly have been some success stories in the fight against infectious diseases. Precisely because they seldom hit the headlines, I would like to mention a few of them here:

Fight against river blindness

In 1974 eleven West African states initiated a program to combat river blindness, caused by the onchocerca worm. An onslaught against the flies that transmit river blindness was launched in an area with 20 million inhabitants, two million of whom were infected with the onchocerca worm and some 200,000 of whom were already blinded. Insecticides were incessantly sprayed onto standing bodies of water from helicopters. More than 600,000 cases were prevented, while those who had already contracted the disease were treated. The program costs 600 million dollars, less than one dollar per protected person per year. It will yield an estimated 3.7 billion dollars in savings, because the inhabitants can invest their full energy in farming. The program was supported by the World Bank, the WHO, the UN, the governments of the twenty African countries affected, twenty-seven donor countries, more than thirty NGOs and the pharmaceutical company Merck & Co., which provided the medication Ivermectin free of charge.

Antibiotics against blindness

Since 1997 Morocco has been running a campaign against the trachoma pathogen *Chlamydia trachomatis*, the most frequent cause of blindness. Pfizer is providing the necessary antibiotic free of charge, while timely surgery is putting a stop to loss of vision from the outset. In addition, frequent face washing and other hygiene measures are preventing the pathogens from becoming established in the eyes. This, coupled with improved sanitary facilities and the provision of clean drinking water, has achieved sustained success. The incidence of trachoma in children under the age of ten has fallen by 99%.

Dracunculosis vanquished

In the mid-1980s twenty countries in Asia and sub-Saharan Africa took up the fight against the guinea worm. The guinea worm is a filarial nematode that is transmitted via infected water fleas in contaminated drinking water. The worm spreads in subcutaneous tissue and has given the clinical picture its name: dracunculosis, derived from the Latin for "little dragon" illness. To fight the disease a program was developed that totally dispensed with classical medical interventions and went straight for the cause. Deep wells were dug to tap drinking water that was purified by means of simple filters. Elderly men were posted at contaminated pools to warn people of the danger of infection and advise them to draw water from a well. When the fight against the guinea worm began in 1986 there were still some 3.5 million cases of the disease in the region; in 2006 the number had fallen to 25,000. The program was jointly funded by the Carter Center (launched by former US President Jimmy Carter), the Centers for Disease Control of the USA, UNICEF, the WHO, the Bill & Melinda Gates Foundation, the World Bank, the UN, numerous NGOs, fourteen donor countries, private business, and the governments of the twenty affected countries in Asia and Africa. In those twenty years the costs amounted to no more than 225 million dollars.

Millennium villages in Africa

To prove to the world that, with enough willpower, the Millennium Development Goals really can be achieved – and perhaps also out of frustration over the failure so far to do so – Jeffrey D. Sachs, together with the United Nations, but with purely private donations, has begun to build millennium villages throughout Africa. One of the main supporters is George Soros, a Hungarian-born US multibillionaire, investment banker and philanthropist. It is hoped that the villages will demonstrate how much can be attained with little money. The objective is to eradicate misery for no more than seventy-five dollars per person per year. In addition, the organizers plan to wage war against infectious diseases. This will primarily entail basic immunization against childhood diseases, mosquito nets to help fight malaria, education how to prevent HIV infection, AIDS and tuberculosis medication, as well as the elimination of hitherto neglected tropical diseases. The project started in 2006 with a dozen villages. Today there are seventy-eight villages in twelve African countries with a total of 400,000 inhabitants. One thousand villages are planned by 2009. To achieve these goals the villages will be supported bottom up, whereas in fact the Millennium Development Goals have a mark-edly top-down approach.

A key personal experience

Let me briefly outline how I gained personal hope. It was in South Africa, in Cape Town and Stellenbosch to be precise, where after spending some days in the Townships I visited the Desmond Tutu HIV and TB Centres. These facilities are named after the former Archbishop of South Africa and Nobel Peace Laureate Desmond Tutu. They aim to combine scientific work with medical care and social support in rural communities. They want to give something back to the community, especially in villages where scientific projects are being carried out. Young women with HIV, whose lives have become meaningful again thanks to ART, give encouragement and moral support to other sufferers of AIDS or tuberculosis. Through information and elucidation, an understanding for the ill is promoted among the healthy and their stigmatization is combated. Precautionary and protective measures are explained so that young girls are also able to find out how they can protect themselves from infection and learn to take charge of their own bodies.

In one of these centers we met with Zelphina Maposela, known to everyone as Mama Maposela. She herself grew up as an orphan and today has adopted twenty-three orphaned chil-dren herself. During the day she looks after dozens more. In addition, Mama Maposela is involved in the local HIV support group and runs a soup kitchen for the needy. With his formidable powers of persuasion Desmond Tutu motivates the people at the Centres, which have achieved admirable results for people suffer-ing from AIDS and tuberculosis. Tutu is credible because he has no political ties and expresses his opinion freely – and because he himself barely survived tuberculosis at the age of fifteen. With simply worded sentences he inspires and encourages caregivers. "Those of you who work to care for people suffering from AIDS and tuberculosis are wiping a tear from God's eye."

I am deeply moved by projects of this kind. They instill hope and optimism because they show that something *can* be achieved if all concerned – politicians, opinion-makers, foundations, governments, and governmental and nongovernmental organizations – pull together. Such bottom-up approaches should spur us toward greater commitment and exploring new avenues. Problems will never be solved if we keep blaming each other and dodging responsibility.

11.6 Sunny with cloudy spells

Three times infectious diseases have had devastating repercussions. In the mid-14th century the Black Plague wrought havoc in Europe, decimating the population by one-third to one-quarter. After the First World War there was an outbreak of Spanish Flu with approximately 50 million dead. Toward the end of the twentieth century came HIV/AIDS, with a toll of more than 25 million fatalities so far, and an estimated 35 million more infected. So, what will the future bring? Difficult though predictions are, in the following I will develop three virtual scenarios. While two of them admittedly are extreme situations, all of them are realistic.

Worst case
In the worst-case scenario, the major epidemics are not rolled back because poverty continues to grow in developing countries and less and less money is available for healthcare. Although all kinds of organizations do their utmost to get AIDS, tuberculosis,

and malaria under control, the increasing resistance of the pathogens makes their efforts more and more into a labor of Sisyphus. Current patent legislation and minimal business incentives cause the pipelines for new drugs and vaccines to dry up. Up to 2010 only a handful of new medications are being launched, perhaps two against AIDS, one or two against tuberculosis, and one against malaria. In the industrialized world hardly anyone bothers to address the problem of the neglected tropical diseases. The low level of financial support for research is partly to blame for the lack of effective vaccines. By 2012 extremely resistant tuberculosis has spread to all continents, and an HIV/AIDS tsunami has swept across China, India, and Russia.

The low interest in developing new anti-infectives, and the careless use of antibiotics in hospitals as well as in industrialized livestock breeding make the proportion of resistant bacteria skyrocket. Many diseases become untreatable. Mankind falls back into the pre-antibiotic era of the first half of the past century.

National interests increasingly weaken the WHO, and the development of a global monitoring and surveillance system for the early identification and control of new pathogens cannot get off the ground. Local outbreaks of H5N1 adapted to humans develop into a pandemic. The development of a vaccine against this pandemic influenza progresses sluggishly, with production capacity barely sufficient to meet the needs of the wealthiest countries. In most cases the developing world ends up empty-handed. On top of all this a new pathogen enters the scene. It is disseminated aerially, but like HIV is hardly evident at first. Epidemics take millions of lives – not just in the developing countries but also in the industrialized world.

The economies of China, India and Russia collapse, sending colossal shockwaves through the global economy and resulting in a severe recession. The industrialized countries make drastic cuts in their development aid. The vicious circle closes and the spiraling hardship and suffering drive ever more people from Africa to flee to Europe.

Best case

A virtual best-case scenario should kindle hope, but at the same time be realistic. The prerequisites for this scenario are achievement of the Millennium Development Goals by 2015 as planned and the absence of a pandemic. Moreover, it presupposes a number of innovative approaches to epidemic control and a major change in human behavior. Comprehensive debt relief in exchange for better healthcare roll back vaccine-preventable infectious diseases worldwide. Medical and technological prevention measures (mosquito nets, insecticides, clean drinking water) make substantial inroads against malaria, diarrheal diseases, and many neglected tropical diseases. New financial incentives and a revamped patent system result in the development of new vaccines and drugs. The Agreement on Trade-Related Aspects of Intellectual Property Rights (TRIPS) is amended in such a way as to guarantee that vital drugs and vaccines are available at affordable prices in all countries. In medical practice and animal breeding antibiotics are administered prudently, and the problem of resistance has come largely under control.

Increasingly, anti-infectives, even newly developed ones, are made available to developing countries free of charge. Together with improved preventive measures, in particular thanks to an increasing acceptance of condoms and new microbicide gels, the spread of AIDS is halted and in some countries the disease is even retreating.

The WHO's international role is strengthened and, under its leadership, a well-functioning global monitoring and surveillance system is developed that prevents the spread of an H5N1 that is dangerous to humans. Success is achieved not only thanks to the WHO: numerous other nongovernmental and governmental organizations as well as PPPs also contribute to a substantial extent. They are supported by charismatic politicians and wealthy sponsors who have recognized the threat to world peace and the global economy posed by infectious diseases. Endowed with the necessary authority, national health agencies, working closely with the WHO, rapidly come up with solutions to the problems of global health policy.

The state of global health is greatly improved – partly for humanitarian reasons, partly out of self-interest. It all pays off quickly, for the economies of the developing countries start to grow once they have been freed from the double yoke of poverty and illness. Increasingly, education in poor countries improves – in equal measure for boys and girls. Environmental policy is underpinned and there is a stronger emphasis on conserving nature. Slowly but surely, the developing countries emerge as equal trading partners of the industrialized countries, freeing up umpteen billions of additional euros every year. The economic upswing goes hand in hand with a strengthening of democracy, as the firm belief begins to hold sway that sound health is a basic prerequisite for a life in freedom and self-determination.

Minimum requirements – not easy, but doable

Microbes and infectious diseases grow exponentially, so if we do not start to change our way of thinking now and keep muddling through the way we have been, our slide down into the worst-case scenario could pick up speed very quickly. So, what are the minimum requirements we must meet that are realistically achievable? Whatever happens, the change in thinking is going to have to begin this decade – both in the way we combat the causes and the way we treat the symptoms. We urgently need new technologies, new drugs, new vaccines. This can only happen if networked associations of research institutes, especially ones modeled on the PPP principle, receive sufficient incentives through selective promotion. Amended patent legislation and easier global access to new technologies, drugs, and vaccines will make the fruits of this research affordable to poor countries as well. A reduction of poverty and improvement of healthcare through an exchange of debt for health programs with the involvement of the most successful aid organizations can begin to break the vicious circle between poverty and the major epidemics. We all have to change our behavior and insist that antibiotics are administered prudently, that animal husbandry is carried out on a less intensive scale, and that a large part of our foodstuffs is produced locally and not shifted back and forth across continents and oceans. All this greatly diminishes the risk of a new epidemic developing. Although the industrialized countries will focus their efforts on combating those infectious diseases to which they are most vulnerable, the intensive research in this field will release forces that will also drive forward the successful development of intervention strategies against the neglected diseases.

To avoid outbreaks of new epidemics, an effective globally interlinked monitoring and surveillance system must be built up that immediately identifies newly emergent pathogens and deploys rapid-response mechanisms to stop them spreading around the globe. Coordinated by the WHO, international health regulations are considerably strengthened to promote the development of early warning systems.

11.7 Outlook

Even the minimum solution demands that we show commitment.
I hope this book will inspire you to think about the problem,
discuss it with your friends and colleagues, urge politicians to
take action, support PPPs, governmental and nongovernmental
organizations, and demand that the pharmaceutical industry
and scientific institutions come up with solutions.

After the end of the Second World War – on July 5, 1947,
to be precise – the Marshall Plan was launched. This aid
program borne by the USA made a decisive contribution to the
reconstruction of the shattered and impoverished countries of
Western Europe. Between 1947 and 1951 America gave 1% of
its gross national income to fund it. In those days that was just
under 13 billion dollars, today it would amount to roughly 100
billion dollars. A new Marshall Plan for the developing countries
in which all industrialized countries participate would not only
bring epidemics under control but would also achieve other Millennium Development Goals. I think we would be well advised
to make a committed effort in this direction. To conclude, here
is a frequently quoted *bon mot* from Voltaire that is very much
to the point:

"We are responsible for what we do – but also for what we don't
do."

Glossary

Acquired immunity: The body's reaction to foreign substances, mediated by B cells and T cells and characterized by specificity and memory.

Acquired immunodeficiency syndrome (AIDS): One of the three major infectious diseases. Around 35 million people are now infected with the causative agent, human immunodeficiency virus (HIV). The virus weakens the immune system. As a result, infected people develop secondary infections and, if untreated, die.

Acute infection: Infection with a pathogen that causes illness within a few days.

Anthrax: A bacterial infectious disease that is commonly fatal. The first infectious disease shown (in 1876 by Robert Koch) to be due to a specific microorganism. Now feared as a potential biological weapon.

Antibiotic: A drug used to treat infection with microorganisms, in particular bacteria.

Antibody: A protein present in the blood that specifically recognizes antigens. Antibodies can combat, among other things, pathogens that have gained entry to the body.

Antigen: A structure that is specifically recognized as foreign by the body's immune system and that triggers an immune response.

Anti-infectives: A generic term for various substances used to combat pathogens; includes antibiotics, chemotherapeutic agents, and vaccines.

Antiretroviral therapy (ART): Treatment of AIDS with one or more drugs.

Antituberculous drugs (tuberculostatics): Drugs used to treat tuberculosis.

Artemisinin: A new plant-derived drug for the treatment of malaria.

Attenuation: Weakening of a pathogen so that it loses its ability to cause illness. Important for vaccine production.

Avian influenza: A disease of birds caused by influenza viruses. H5N1, the best-known representative, can become dangerous to humans after undergoing genetic change.

B cells: One of the two lymphocyte populations present in blood; responsible for the production of antibodies.

Bacterium (plural: *bacteria*): A type of unicellular microorganism. Bacteria are found in every biological niche on Earth. Some live in or on human beings and sometimes cause disease.

Bed net: A fine net used to protect against diseases, in particular malaria, that are transmitted by insects. Nowadays mostly impregnated with insecticides.

Blockbuster: A drug with annual sales of over one billion dollars.

Bovine spongiform encephalopathy (BSE): A disease of cattle caused by prions. Though not caused by a living creature, BSE is transmissible. See also *Prion*.

Campylobacter jejuni: A common bacterial cause of diarrheal disease and food poisoning.

CD4 T cells: An important population of T lymphocytes, in particular T helper cells, which activate functions in other cells by means of soluble mediators.

CD8 T cells: An important population of T lymphocytes that typically destroy infected cells (= T killer cells).

Cervical carcinoma: Cancer of the cervix (neck) of the uterus (womb). Caused by human papillomaviruses.

Chemotherapy: Drug therapy of illnesses with antibiotics, antiviral substances, etc.

Chickenpox: A childhood illness caused by a virus.

Cholera: A transmissible disease marked by watery diarrhea which if untreated can cause death from dehydration.

Chronic carrier: A person who bears and excretes a pathogen without being ill.

Chronic infection: Infection with a pathogen that can persist for years and sometimes causes illness.

Cold: Popular term for a group of mostly harmless infectious diseases characterized by cough, nasal congestion, and sore throat.

Commensalism: (here) A form of coexistence between microorganisms and humans or animals that does not cause harm to either.

Commission on Macroeconomics and Health: A commission established by the World Health Organization to develop economic solutions to major health problems including infectious diseases.

Complement: A group of soluble immune substances present in the blood involved in antimicrobial defense.

Compulsory licensing: Lifting of patent law to make lifesaving drugs available at an affordable price, e.g. to treat AIDS in developing countries.

Condom: A protective sheath applied to the penis to prevent conception and to protect against sexually transmitted diseases. At present condoms are the cheapest and most effective preventive measure against HIV/AIDS.

Cytokine: A soluble mediator that conveys information between cells of the immune system.

DDT (dichlorodiphenyltrichloroethane): A disinfectant and insecticide that plays an important role in the fight against malaria and other vector-borne diseases.

Dengue: A viral disease transmitted by mosquitoes; causes a hemorrhagic fever that is often fatal.

Developing country: A country at an early stage of economic and political development.

Disability-adjusted life year (DALY): A measure that represents one year of healthy life lost. An important parameter for cost-benefit analyses of diseases.

DNA (deoxyribonucleic acid): A substance that bears the genetic information of all living creatures. See also *RNA*.

Doha Round: A series of meetings of the World Trade Organization on problems that world trade causes in developing countries. Among other things, it aimed to permit patent-free production and sale of lifesaving medicines at an affordable price in poor countries. In 2006 negotiations broke down before any definite agreement had been reached.

DOTS (directly observed treatment, short course): A strategy for treating tuberculosis in which a combination of drugs is taken under the direct supervision of medical or paramedical personnel.

Dysentery: Diarrheal disease due to bacteria or parasites.

Ebola: A viral disease transmitted via blood that causes bleeding throughout the body.

Endemic: The presence of a disease in a certain area.

Epidemic: A spatially and temporally limited outbreak of a disease.

Epidemiology: The study of epidemics, pandemics, and endemics.

Escherichia coli: A normal bacterial inhabitant of the intestine that has become the "workhorse" of molecular biological research.

Expanded Program on Immunization (EPI): An international program for primary immunization of children.

G8: A group composed of the seven most powerful industrialized countries (Germany, France, the United Kingdom, Italy, Japan, Canada, and the USA) plus Russia that accounts for two-thirds of world trade and gross world income but less than 15% of the world's population.

Global Alliance for Vaccines and Immunization (GAVI = GAVI Alliance): An international organization for carrying out immunization programs in developing countries.

Global Fund to Fight AIDS, Tuberculosis and Malaria (GFATM): An organization set up in the year 2000 to combat the major infectious diseases.

GO: Governmental organization.

H5N1: see *Avian influenza*.

HAART (highly active antiretroviral therapy): Extended combination therapy for AIDS. See also *ART*.

Health: "A state of complete physical, mental, and social wellbeing and not merely the absence of disease or infirmity," according to the World Health Organization's definition.

Helicobacter pylori: A bacterium that can cause gastritis, stomach ulcers, and stomach cancer.

Helminths: Multicellular organisms (worms) with a complex lifecycle. Larvae parasitic in insects or snails, adult forms in humans; responsible for many neglected tropical diseases.

Hemorrhage: (here) Bleeding in a number of organs brought about by viral pathogens.

Hepatitis: Inflammation of the liver due to various hepatitis viruses that cause illness of varying degrees of severity.

Host range: The range of animals, including man, that a pathogen can infect.

Host: (here) A mammal, including man, that can be infected and made ill by pathogens.

Human immunodeficiency virus (HIV): see *AIDS*.

Human papillomavirus: A virus responsible for cervical cancer in women.

Hygiene: Measures taken to prevent spread of infectious diseases.

Immune response: The body's defense reaction against pathogens.

Immune system: The body's defense system. Consists of a nonspecific innate system and an antigen-specific acquired system.

Immunity: The ability of the body to take specific action against pathogens by means of an immune response.

Immunization: Also termed vaccination. Inoculation of attenuated pathogens or components of pathogens (= vaccines) in order to induce specific immunity; an important preventive measure against many transmissible diseases; the most cost-effective measure in medicine.

Immunoglobulin (Ig): see *Antibody*.

Incidence: The number of new cases of a disease occurring within a certain period of time (generally a year).

Incubation period: The period between entry of a pathogen into its host and the onset of illness. See also *Acute infection* and *Chronic infection*.

Industrialized country: A country with a high level of technical and economic development.

Infection: A conflict between microbial pathogens and our body; may, but does not necessarily, lead to an infectious disease.

Infectious disease: A disease that results from infection.

Inflammation: The body's reaction to injury, infection, and the like, characterized by accumulation of tissue fluid and blood cells.

Influenza virus: see *Influenza, Avian influenza, H5N1*.

Influenza: A globally important disease caused by influenza viruses that can occur in epidemic or pandemic form.

Interferon gamma: Important soluble mediator produced by T helper cells of type 1 (Th1 cells); activates defense mechanisms in macrophages.

Interleukin: A generic term for a group of soluble mediators produced by T helper cells that mediate the immune response.

International Health Regulations: A set of regulations for global control of infectious diseases coordinated by the World Health Organization; revised in mid-2007.

Latent infection: Peaceful coexistence of a pathogen with its host; may, but does not necessarily, lead to illness.

Live vaccine: A vaccine derived from an attenuated live pathogen.

Lymphocyte: A type of white blood cell that is responsible for acquired immunity. There are two types: B lymphocytes (B cells) and T lymphocytes (T cells).

Macrophage: A cell that fights bacterial infections by engulfing (ingesting) bacteria.

Malaria: A tropical disease that is caused by protozoa (malaria plasmodia) and transmitted by insects (anopheles mosquitoes).

Marburg virus: A virus that causes a rare form of hemorrhagic fever with a high mortality rate.

Marshall Plan: An aid program instituted by the USA after the Second World War to help reconstruct the ravaged countries of Western Europe by means of credits, food, and goods to a value of 13 billion dollars (equivalent to 100 billion dollars today).

MDR-TB (multidrug-resistant tuberculosis): A form of tuberculosis against which the best drugs are no longer effective. See also *Tuberculosis*.

Measles: A viral infectious disease for which a vaccine is available.

Médecins Sans Frontières (Doctors Without Borders): An international organization that provides medical assistance in problem areas. Recently it has also helped in the struggle against infectious diseases. It was awarded the Nobel Peace Prize in 1990.

Meningitis: Inflammation of the membranes that surround the brain. Can be caused by bacteria or viruses.

Microbe: see *Microorganism*.

Microbicide: A substance that kills microbes. Microbicides that kill HIV in the vagina would empower women to protect themselves against HIV infection.

Microorganism: A microscopically small living creature, generally unicellular (bacteria, fungi, protozoa, viruses).

Millennium Development Goals: An undertaking made at the 55th General Assembly of the United Nations in 2000 to eradicate or at least reduce extreme poverty, hunger, child mortality, and major infectious diseases and to improve school education, gender equality, and environmental sustainability by the year 2015. It is to be feared that most of these goals will not be achieved. See also *United Nations*.

Morbidity: The proportion of ill people in a population. Here: a measure of the likelihood of suffering from a particular infectious disease.

Mortality: The proportion of people in a population who die. Here: a measure of the likelihood of dying from a particular infectious disease.

Mumps: A viral disease against which a vaccine is available.

Neglected tropical diseases: Infectious diseases that are rife in the tropics but generally ignored in industrialized countries. Most are caused by protozoa and worms.

NGO: Non-governmental organization.

Normal flora: The totality of microorganisms that are naturally present in a certain part of the body.

Nosocomial infection: An infectious disease that is transmitted in hospitals, old-age homes, and similar institutions. See also *Opportunist*.

OECD: Organization for Economic Co-operation and Development. An international organization made up largely of industrialized countries that promotes world trade and economic growth. The OECD presently provides more than 100 billion dollars worth of assistance to developing countries.

Opportunist: A microorganism that normally does not cause illness in healthy individuals but may do so in debilitated individuals.

Orphan Drug Act: US and EU law to promote the development of drugs for use against very rare diseases.

Pandemic: A temporally limited but worldwide outbreak of a disease.

Parasite: An organism that obtains its nutrition from a living host.

Pathogen: A microorganism that causes illness.

Pathogenesis: The mechanism by which a disease develops.

Pathogenicity: Ability to cause disease.

Phagocyte: A cell that can engulf (ingest) particulate matter. See also *macrophage*.

Plasmodia (singular: *plasmodium*): A group of protozoa that includes the causative agent of malaria.

Pneumonia: Inflammation of the lungs.

Poliomyelitis: A viral disease against which a vaccine is available. Also known as polio.

PPP: Public-private partnership.

Prevalence: The absolute number of cases of a given disease in a population, including both newly acquired disease and disease that has been present for some time.

Prion: An altered protein present in the brain that is responsible for bovine spongiform encephalopathy in cattle and variant Creutzfeldt-Jakob disease in humans.

Protozoa (singular: *protozoon*): A group of unicellular higher organisms, some of which cause disease.

Quarantine: Isolation of highly infectious humans or animals that represent a danger to the general population.

RNA (ribonucleic acid): A substance that transmits genetic information. See also *DNA*.

Rotavirus: A group of viruses that cause diarrheal disease, especially in young children, against which a vaccine has recently become available.

Rubella (German measles): A highly infectious disease that can cause serious complications in pregnant women. A vaccine is available.

*Salmonellae (*singular: *salmonella)*: A group of bacteria transmitted via food that can cause diarrheal disease and typhoid fever, among other illnesses.

SARS (severe acute respiratory syndrome): A type of viral pneumonia that threatened to develop into a pandemic in 2003.

Sepsis: An infection of the circulating blood that is often fatal.

Septic shock: A common and mostly fatal consequence of sepsis. Results from an excessive immune reaction (cytokine storm). See also cytokine.

*Shigellae (*singular: *shigella)*: A group of bacteria that cause severe diarrheal disease.

Smallpox: A viral disease that was eradicated in 1980.

Subunit vaccine: A vaccine consisting of partially purified or pure components of a pathogen.

*Staphylococci (*singular: *staphylococcus)*: A group of bacteria that cause purulent infections and certain forms of food poisoning.

*Streptococci (*singular: *streptococcus)*: A group of bacteria that cause purulent infections and scarlet fever and that if untreated can give rise to rheumatic fever.

Surveillance: (here) Collection, analysis, and interpretation of data on an outbreak of disease.

Symbiont: (here) A microorganism that forms a mutually beneficial association with humans or animals.

T helper cells: see also *CD4 T cells.*

– Type 1 (Th1 cells): produce soluble mediators that activate macrophages and are important for combating bacterial and viral infections and also in autoimmune diseases.

– Type 2 (Th2 cells): produce mediators that are important for combating worm infestations and also in allergic reactions.

T killer cells: see *CD8 T cells.*

Tetanus: A disease due to a bacterial toxin that is characterized by paralysis and is often fatal. A vaccine is available.

Toxoid vaccine: A vaccine consisting of a modified microbial toxin that induces an immune response but no longer acts as a toxin.

Transmissible disease: see *Infectious disease.*

TRIPS (Agreement on Trade-Related Aspects of Intellectual Property Rights): An agreement that covers, among other things, patent protection for drugs and vaccines for use against the major infectious diseases, especially AIDS. At the same time, it entitles states to take measures to protect the health of their population, e.g. by compulsory licensing of essential drugs.

Tuberculosis (TB): One of the major infectious diseases. Caused by a bacterium, generally affects the lungs. The presently available vaccine is effective in children but not in adults.

Type I interferon: A group of soluble immune mediators with antiviral activity.

Typhoid fever: A bacterial disease characterized by diarrhea and severe bouts of fever that is often fatal. Caused by salmonellae.

United Nations (UN): An international organization with 192 member states dedicated to maintaining world peace and promoting human rights and international cooperation.

Vaccine: see *Immunization.*

Vector: An organism that can transmit a pathogen to humans or animals. Important vectors are insects and snails.

Virulence: A measure of the illness-inducing properties of a microorganism.

Virus: A microorganism that is invisible under the light microscope and cannot reproduce by itself. Reproduction of it by host cells often results in illness. Contains either DNA or RNA.

World Bank: A bank originally set up to assist countries affected by the Second World War. Promotes the economic development of developing countries.

World Health Organization (WHO): The United Nations authority for health.

World Trade Organization (WTO): An international organization for trade and economic relations. Currently has 150 members.

XDR-TB (extensively drug-resistant tuberculosis): Tuberculosis due to mycobacteria that are resistant to virtually all antituberculous drugs. Now being found with increasing frequency.

Zoonosis: An infectious disease that is transmitted from animals to humans. The great majority of newly occurring infectious diseases are zoonoses.

References

Chapters 2 to 7

Diamond, Jared (2005), *Guns, Germs, and Steel: The Fates of Human Societies*. New York: WW Norton & Co.

Ewald, Paul W. (2000), *Plague Time: How Stealth Infections Cause Cancers, Heart Disease, and Other Deadly Ailments*. New York: Free Press.

Garrett, Laurie (2000), *Betrayal of Trust: The Collapse of Global Public Health*. New York: Hyperion.

Greger, Michael (2006), *Bird Flu: A Virus of Our Own Hatching*. New York: Lantern Books.

Horton, Richard (2004), *MMR: Science and Fiction. Exploring the Vaccine Crisis*. London: Granta Books.

Hotez, Peter J (2008), *Forgotten People, Forgotten Diseases*. Washington D.C: ASM Press.

Jamison, Dean/Breman, Joel/Measham, Anthony/Alleyne, George/Claeson, Mariam/Evans, David/Jha, Prabhat/Mills, Anne/Musgrove, Philip (2006), eds., *Priorities in Health*. The World Bank.

Jamison, Dean/Breman, Joel/Measham, Anthony/Alleyne, George/Claeson, Mariam/Evans, David/Jha, Prabhat/Mills, Anne/Musgrove, Philip (2006), eds., *Disease Control Priorities in Developing Countries* (2e). The World Bank.

Murphy, Kenneth/Travers, Paul/Walport, Mark (2008),
    Janeway's Immunology, 7th Edition. New York: Garland
    Science.

Kaufmann, S. H. E. (2004), ed., *Novel Vaccination Strategies*.
    Weinheim: Wiley-VCH.

Knight, Lindsay (2008), ed., *World Disasters Report 2008:
    Focus on HIV and AIDS*. International Federation of Red
    Cross and Red Crescent Societies. Satigny/Vernier: ATAR
    Roto Presse.

Lee, Kelley/Collin, Jeff (2005), *Global Change and Health,
    Understanding Public Health*. Maidenhead: Open University
    Press.

Lopez, Alan/Mathers, Colin/Ezzati, Majid /Jamison, Dean/
    Murray, Christopher (2006), eds., *Global Burden of Disease
    and Risk Factors*. New York: Oxford University Press.

Mims, Cedric/Dockrell, Hazel/Goering, Richard/Roitt, Ivan/
    Wakelin, Derek/Zuckermann, Mark (2004), *Medical
    Microbiology*, 3rd Edition. Edinburgh: Elsevier.

National Research Council (2006), *Treating Infectious Diseases
    in a Microbial World. Report of Two Workshops on
    Novel Antimicrobial Therapeutics*. Washington: National
    Academies Press.

Perlin, David/Cohen, Ann (2002), *The Complete Idiot's Guide
    to Dangerous Diseases and Epidemics*. Indianapolis: Alpha
    Books.

Prüss-Üstün, Annette/Corvalán, Carlos (2006), *Preventing
    Disease Through Healthy Environments: Towards an
    Estimate of the Environmental Burden of Disease*. Geneva:
    WHO.

Salyers, Abigail A./Whitt, Dixie D. (2005), *Revenge of the
    Microbes: How Bacterial Resistance is Undermining the
    Antibiotic Miracle*. Washington, DC: ASM Press.

Sherman, Irwin (2006), *The Power of Plagues*. Washington, DC: ASM Press.

The World Bank (2004), *Mini Atlas of Global Development*. Brighton: Myriad Editions.

Beck, Eduard/Mays, Nicholas/Whiteside, Alan/Zuniga, José (2006), eds., *The HIV Pandemic: Local and Global Implications*. New York: Oxford University Press.

Torrey, E. Fuller/Yolken, Robert H. (2005), *Beasts of the Earth: Animals, Humans and Disease*. New Jersey: Rutgers University Press.

WHO (2003), *Emerging Issues in Water and Infectious Disease*. Geneva: WHO.

WHO (2006), *SARS: How a Global Epidemic was Stopped*. Western Pacific Region: WHO.

WHO (2008), *International Travel and Health: Situation as on 1 January 2008*. Geneva: WHO.

Chapters 8 to 11

Africa's missing billions, IANSA, Oxfam, and Saferworld (2007), http://www.oxfam.org/en/files/bp107_africas_missing_billions_ 0710.pdf/download

Amon, Joseph (2006), *Preventing the Further Spread of HIV/AIDS: The Essential Role of Human Rights*. http://hrw.org/wr2k6/hivaids/hivaids.pdf.

Atlas der Globalisierung (2006), *Die neuen Daten und Fakten zur Lage der Welt*. Le Monde diplomatique.

Beah, Ishmael (2007), *A Long Way Gone. Memoirs of a Boy Soldier*. London: Harper Collins.

Richner, Beat (2004), *Hope for the Children of Bopha Kantha*. Zürich: Nzz Libro.

Bowen-Jones, Evan (1998), *A Review of the Commercial Bushmeat Trade with Emphasis on Central/West Africa and the Great Apes*. Cambridge: Ape Alliance.

Broekmans, Jaap/Caines, Karen/Paluzzi, Joan/Sachs, Jeffrey D. (2005), eds., *Investing in strategies to reverse the global incidence of TB*. UN Millennium Project, Task Force on HIV/AIDS, Malaria, TB, and Access to Essential Medicines, Working Group on TB.

Brownlie, Joe/Peckham, Catherine/Waage, Jeffrey/Woolhouse, Mark/Lyall, Catherine/Meagher, Laura/Tait, Joyce/Baylis, Matthew/Nicoll (2006), *Infectious Diseases: Preparing for the Future: Future Threats*. Office of Science and Innovation. London: Department of Trade and Industry.

Corvalan, Carlos/Hales, Simon/McMichael, Anthony (2005), *Ecosystems and Human Well-being: Health Synthesis*. Millennium Ecosystem Assessment. Geneva: WHO.

Delgado, Christopher L./Rosegrant, Mark W./Meijer, Siet/Ahmed, Mahfuzuddin (2003), *Outlook for Fish to 2020: Meeting Global Demand*. Penang, Malaysia: World Fish Center. http://www.ifpri.org/pubs/fpr/pr15.pdf.

Finkelstein, Stan/Temin, Peter (2008), *Reasonable Rx: Solving the Drug Price Crisis*. New Jersey: FT Press.

Food and Agriculture Organization (FAO) of the United Nations (2006), *Global Forest Resources Assessment 2005: Progress Towards Sustainable Forest Management*. Forestry Paper 147. Rome: FAO. http://www.fao.org/forestry/site/fra2005/en/.

Garrett, Laurie (1995), *The Coming Plague: Newly Emerging Diseases in a World Out of Balance*. New York: Penguin.

Garrett, Laurie (2000), *Betrayal of Trust: The Collapse of Global Public Health*. New York: Hyperion.

Garrett, Laurie (2005), HIV and National Security: Where are the Links? A Council on Foreign Relations Report. http://www.casy.org/engdocs/HIV_National_Security.pdf.

Guillemin, Jeanne (2005), *Biological Weapons: From the Invention of State-sponsored Programs to Contemporary Bioterrorism*. New York: Columbia University Press.

Herlihy, David (1997), *The Black Death and the Transformation of the West*. Cambridge, MA: Harvard University Press.

Hochschild, Adam (2000), *King Leopold's Ghost: A Story of Greed, Terror and Heroism in Colonial Africa*. Boston: Mariner Books.

Institute of Medicine (2007), *PEPFAR Implementation: Progress and Promise*. Washington, D.C.: National Academies Press.

International Finance Facility (2005), *The International Finance Facility*. London: HM Treasury.

Juma, Calestous/Yee-Cheong, Lee (2005), *Innovation: applying knowledge in development*. UN Millennium Project, Task Force on Science, Technology, and Innovation. London: Earthscan.

Kaufmann, Stefan H. E. (2003), *Taking Arms Against a Sea of Troubles*. Max Planck Research 4: 15–19.

Kovats, Sari/Ebi, Kristie L./Menne, Beettina (2003), *Methods of Assessing Human Health Vulnerability and Public Health Adaptation to Climate Change*. Health and Global Environmental Change. Series No. 1. Denmark: WHO Europe. http://www.euro.who.int/document/e81923.pdf.

Kuhn, Katrin/Campbell-Lendrum, Diarmid/Haines, Andy/Cox, Jonathan (2005), *Using Climate to Predict Infectious Disease Epidemics*. Geneva: WHO. http://www.who.int/globalchange/publications/infectdiseases.pdf.

Leach, Beryl/Palluzi, Joan/Munderi, Paula (2005), *Prescription for Healthy Development: Increasing Access to Medicines*. UN Millennium Project, Task Force on HIV/AIDS, Malaria, TB, and Access to Essential Medicines, Working Group on Access to Essential Medicines. London: Earthscan.

Levine, Ruth (2007), *Case Studies in Global Health: Millions Saved*. Sudbury MA: Jones and Bartlett.

Lomborg, Bjorn (2004), *Global Crises, Global Solutions*. Cambridge: Cambridge University Press.

Lomborg, Bjorn (2006), *How to Spend $ 50 Billion to Make the World a Better Place*. Cambridge: Cambridge University Press.

Macdonald, Théodore H (2007), *The Global Human Right to Health: Dream or Possibility?* Oxford: Radcliffe Publishing.

Martens, Jens/Hain, Roland (2002), *Globale öffentliche Güter: Zukunftskonzept für die internationale Zusammenarbeit*. Working Paper. Berlin: Heinrich-Böll-Stiftung.

Nestle, Marion (2003), *Safe Food: Bacteria, Biotechnology, and Bioterrorism*. Berkeley, Los Angeles: University of California Press.

Nierenberg, Danielle (2005), *Happier Meals: Rethinking the Global Meat Industry*. Worldwatch Paper 171. Washington, DC: Worldwatch Institute.

Noah, Don/Fidas, George (2000), eds., *National Intelligence Council Special Reports. National Intelligence Estimate: The Global Infectious Disease Threat and Its Implications for the United States. Environmental Change and Security Project Report*, Issue 6 (Summer 2000). http://www.wilsoncenter.org/topics/pubs/Report6-3.pdf.

O'Donovan, Diarmuid (2008), *The Atlas of Health: Mapping the Challenges*. London: Earthscan.

Preston, Richard (1995), *The Hot Zone*. New York Anchor Books Doubleday.

Ruxin, Josh/Binagwaho, Agnes/Wilson, Paul (2005), *Combating AIDS in the Developing World*. UN Millennium Project, Task Force on HIV/AIDS, Malaria, TB, and Access to Essential Medicines, Working Group on HIV/AIDS. London: Earthscan.

Sachs, Jeffrey D. (2001), *Macroeconomics and Health: Investing in Health for Economic Development*, Report of the Commission on Macroeconomics and Health. Geneva: WHO.

Sachs, Jeffrey D. (2005), *The End of Poverty. Economic Possibilities for our Times*. USA: Penguin Group.

Saker, Lance/Lee, Kelley/Cannito, Barbara/Gilmore, Anna/Campbell-Lendrum, Diarmid (2004), *Globalization and Infectious Diseases: A Review of the Linkages*. Social, Economic and Behavioural (SEB) Research. Special Topics No. 3. Geneva: WHO. http://www.who.int/tdr/cd_publications/pdf/seb_topic3.pdf.

Shah, Sonia (2007), *The Body Hunters: Testing New Drugs on the World's Poorest Patients*. New York: The New Press.

Skolnik, Richard (2008), *Essentials of Global Health*. Boston: Jones and Bartlett Publishers.

Spiegel Special (2007), *Afrika: Das umkämpfte Paradies*. Nr. 2. 2007.

Sutherst, Robert (2004), "Global change and human vulnerability to vector-borne diseases". *Clinical Microbiology Reviews* Vol. 17: 136–173.

Teklehaimanot, Awash/Singer, Burt/Spielman, Andrew/Tozan, Yeim/Schapira, Allan (2005), *Coming to Grips with Malaria in the New Millennium*. UN Millennium Project, Task

Force on HIV/AIDS, Malaria, TB and access to essential
medicines, working group on Malaria. London: Earthscan.

The Data Report 2007 (2007), http://www.thedatareport.org/.

The Earth Institute (2007), Annual Report for Year 1 Activities:
February 2006–February 2007. Millennium Research
Villages. New York: Columbia University, The Earth
Institute.

The World Bank (2005), World Development Report 2006:
Equity and Development. Washington, D.C.: World Bank
Publications.

The World Bank (2006), World Development Report 2007:
Development and the Next Generation. Washington, DC:
The World Bank.

Transparency International (2006), Global Corruption Report
2006, Special Focus, Corruption and Health. London: Pluto
Press.

UNAIDS (2004), Debt-for-AIDS Swaps: A UNAIDS Policy
Information Brief. UNAIDS/04.13E. Geneva: WHO.
http://data.unaids.org/Publications/IRC-pub06/JC1020-
Debt4AIDS_en.pdf

UNAIDS/UNHCR (2005), Strategies to Support the HIV-
related Needs of Refugees and Host Populations. Geneva:
UNAIDS. http://data.unaids.org/publications/irc-pub06/
JC1157-Refugees_en.pdf.

WHO (2008), The World Health Report 2008: Primary Health
Care: Now More Than Ever. Geneva: WHO.

Important online resources and links

AIDS, malaria, tuberculosis

http://academic.sun.ac.za/tb/ (Desmond Tutu TB Centre, Stellenbosch, South Africa)

http://www.desmondtutuhivcentre.org.za/ (Desmond Tutu HIV Centre, Cape Town, South Africa)

http://www.unaids.org/en/ (Joint United Nations Programme on HIV/AIDS, UNAIDS, Geneva, Switzerland)

http://www.globalaidsalliance.org/ (Washington, DC, USA)

http://www.pepfar.gov/ (The US President's Emergency Plan for AIDS Relief, PEPFAR)

http://www.fightingmalaria.gov/ (President's Malaria Initiative, Washington, DC, USA)

http://www.rbm.who.int/ (The Rollback Malaria Partnership, Geneva, Switzerland)

http://www.stoptb.org/ (STOP TB, Geneva, Switzerland)

Infectious diseases, epidemics

http://www.beat-richner.ch/ (Dr. med Beat Richner, Kantha Bopha Children's Hospital)

http://www.bt.cdc.gov/ (Bioterrorism site of Centers for Disease Control and Prevention, CDC, Atlanta, GA, USA)

http://www.cdc.gov/ (Centers for Disease Control and Prevention, CDC, Atlanta, GA, USA)

http://www.cidrap.umn.edu/ (Center for Infectious Disease Research and Policy, University of Minnesota, Minneapolis, MN, USA)

http://www.ecdc.eu.int/ (European Centre for Disease Prevention and Control, ECDC, Stockholm, Sweden)

http://www.wpro.who.int/sites/epi/ (Expanded Program Immunization of the WHO, EPI)

http://www.mpiib-berlin.mpg.de/ (Max Planck Institute for
   Infection Biology, Berlin, Germany)
http://www.nih.gov/ (National Institutes of Health, Bethesda,
   Maryland, USA)
http://www.niaid.nih.gov/ (National Institutes of Allergy and
   Infectious Diseases, Bethesda, Maryland, USA)
http://www.rki.de/ (Robert Koch Institut, Berlin, Germany)
http://www.aerzte-ohne-grenzen.de/ (Médècins Sans Frontieres)

Foundations, PPPs

http://www.theglobalfund.org/en/ (The Global Fund to Fight
   AIDS, Tuberculosis and Malaria, Geneva, Switzerland)
http://www.gavialliance.org/ (Gavi Alliance; formerly Global
   Alliance for Vaccination and Immunisation, Geneva,
   Switzerland)
http://www.gatesfoundation.org/default.htm (Bill & Melinda
   Gates Foundation, Washington, DC, USA)
http://www.wellcome.ac.uk/ (The Wellcome Trust, London,
   UK)

Global goods, human rights

http://www.cgdev.org/ (Center for Global Development,
   Washington, DC, USA)
http://www.cfr.org/ (Council on Foreign Relations, New York,
   USA)
http://www.globalissues.org (Global Issues: Social, Political,
   Economic and Environmental Issues That Affect Us All)
http://www.hrw.org/ (Human Rights Watch, New York, USA)
http://www.crisisgroup.org (International Crisis Group,
   Brussels, Belgium)
http://www.transparency.org/ (Transparency International,
   Berlin, Germany)

http://www.worldwatch.org/ (Worldwatch Institute, Washington, DC, USA)

Economics and trade

http://www.data.org/ (Debt AIDS Trade Africa, Washington, DC, USA)

http://www.wto.org/ (World Trade Organization, WTO, Geneva, Switzerland)

http://www.oecd.org (Organisation for Economic Cooperation and Development, OECD, Paris, France)

http://www.dti.gov.uk/ (Department of Trade and Industry, London, UK)

http://www.copenhagenconsensus.com (Copenhagen Consensus Center, Frederiksberg, Denmark)

United Nations and its organizations

http://www.un.org/ (United Nations, UN, New York, USA)

http://www.fao.org/ (Food and Agriculture Organization of the United Nations, FAO, Rome, Italy)

http://www.unicef.com/ (United Nations Children's Fund, UNICEF, New York, USA)

http://www.un.org/millenniumgoals/ (United Nations, UN, New York, USA)

http://www.who.int/en/ (World Health Organization, WHO, Geneva, Switzerland)

http://www.worldbank.org/ (World Bank, Washington, DC, USA)

Illustrations

All illustrations: Peter Palm, Berlin and Diane Schad, Berlin.